J. Milner Fothergill

Vaso-Renal Change Versus Bright's Disease

J. Milner Fothergill

Vaso-Renal Change Versus Bright's Disease

ISBN/EAN: 9783337242855

Printed in Europe, USA, Canada, Australia, Japan

Cover: Foto ©berggeist007 / pixelio.de

More available books at **www.hansebooks.com**

VASO-RENAL CHANGE;

OR,

SIXTY YEARS' FURTHER STUDY OF THE PATHOLOGICAL PROCESS

FIRST OBSERVED BY

DR. RICHARD BRIGHT, F.R.S.

VASO-RENAL CHANGE

VERSUS

BRIGHT'S DISEASE.

BY

J. MILNER FOTHERGILL, M.D. Edin.,

PHYSICIAN TO THE CITY OF LONDON HOSPITAL FOR DISEASES OF THE CHEST
(VICTORIA PARK);
LATE ASSISTANT-PHYSICIAN TO THE WEST LONDON HOSPITAL;
HON. M.D., RUSH, ILL.;
FOREIGN ASSOCIATE FELLOW OF THE COLLEGE OF PHYSICIANS, PHILADELPHIA.

'Je vieillis.'

NEW YORK: G. P. PUTNAM'S SONS,
27 AND 29, WEST 23D ST.
LONDON: BAILLIÈRE, TINDALL, AND COX.
1887.

'Old age is not an entity, but a set of conditions predisposing to that state which is called Chronic Bright's Disease. And while to most this comes in natural order when the prime of life is run, yet to some old age is no matter of years and averages, but the running down of a spring set for an individual.'

GOODHART.

To the Memory

OF

RICHARD BRIGHT, M.D., F.R.S.,

THE FIRST OBSERVER OF THIS CHANGE,

THIS WORK

IS

Respectfully Dedicated

BY

THE AUTHOR.

PREFACE.

This work is written under a deep sense of responsibility. Of the necessity for some other term than 'Bright's Disease' in the present state of our knowledge, no question can exist. It is now fifteen years since the writer attempted to grapple with this widespread morbid change, in the chapter 'Combined Heart and Kidney Disease,' in the first edition of his work, *The Heart and its Diseases*. This may form some explanation for his venturing upon the present essay. Before taking action, however, he obtained the sanction of the leading professors and lecturers on the Practice of Medicine to the attempt being made; and he has done his best.

His best thanks are due to Dr. Mott, his draughtsman, and Mr. Hanlon, his engraver, for the excellent woodcuts which illustrate the text. Also to Dr. D. G. L. Johnston and Henry T. Wharton, M.A., for their aid in revising the proof sheets.

3, Henrietta Street, Cavendish Square,
June 1, 1887.

LIST OF ILLUSTRATIONS.

FIG.		PAGE
1.	Rabbit's Eye (Fœtal)	30
2.	Section of Kidney Cortex (Granular)	34
3.	Section of Thickened Artery	44
4.	Sphygmographic Tracing	45
5.	Deposits of Urate of Soda	50
6.	Teeth (Gouty)	85
7.	Nails (Gouty)	87
8.	Sections of Two Small Arteries	96
9.	Miliary Aneurysms	100
10.	Section of Normal Kidney	103
11.	Schema of Kidney	104
12.	Glomerulus of Kidney (Healthy)	105
13.	Early Change in Glomerulus	111
14.	Section of Granular Cortex	112
15.	,, ,, ,, More Advanced	113
16.	Tube Casts	118
17.	Sections of Arteries	157
18.	Heart Fibres (Fatty)	158
19.	Section of Congested Granular Kidney	171

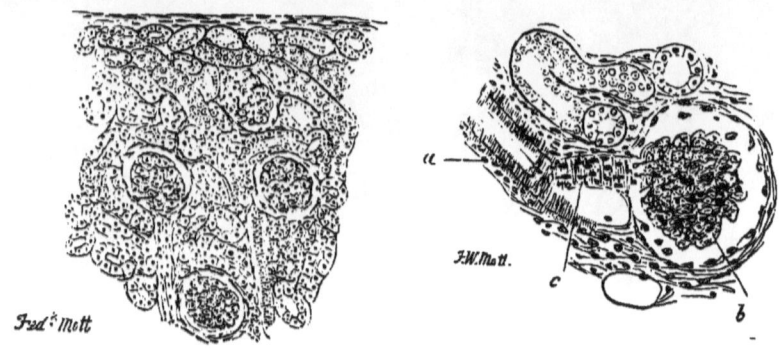

Fig. 10, page 103.

Fig. 13, page 111.

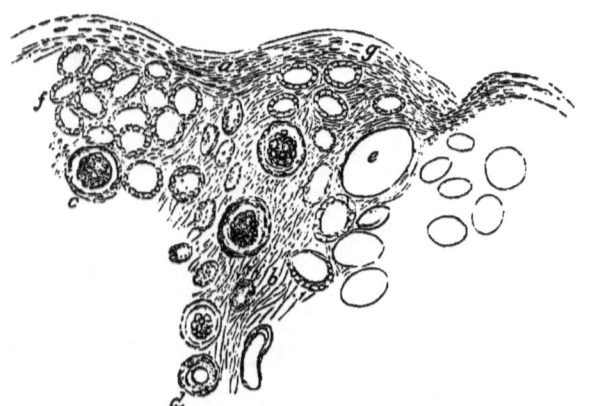

Fig. 2, page 34.

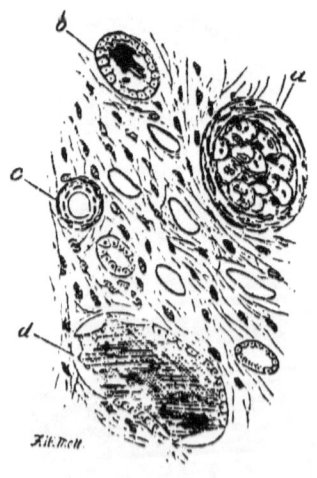

Fig. 15, page 113.

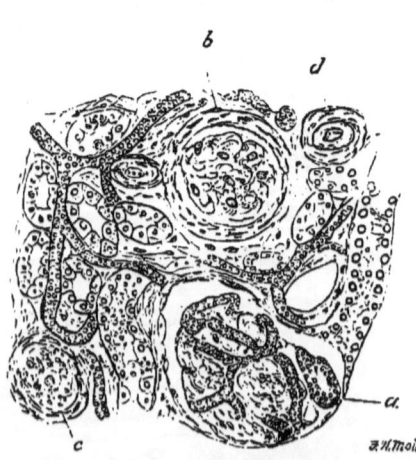

Fig. 19, page 171.

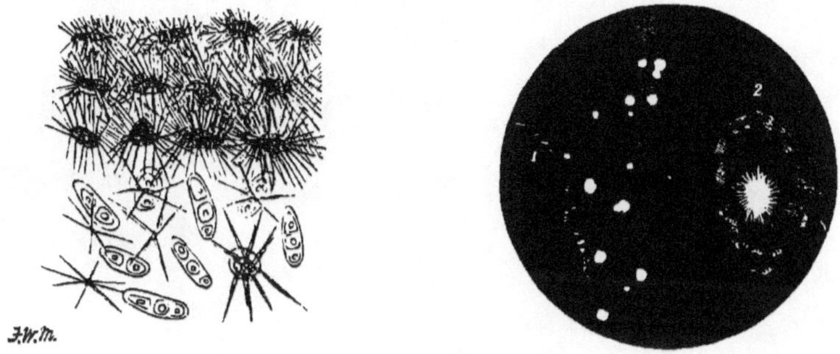

(a) Fig. 5, page 50. (b)

Fig. 3, page 44. 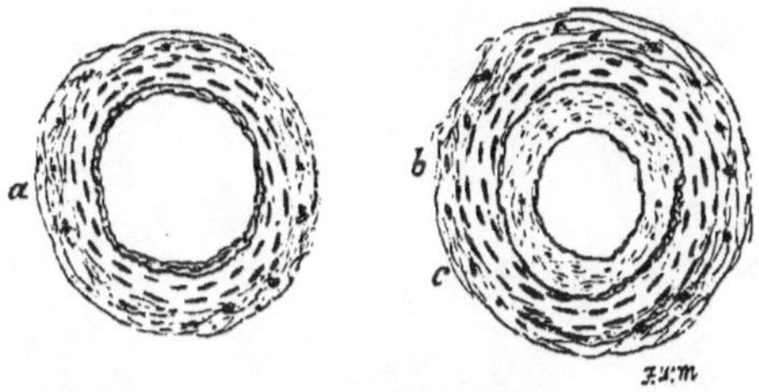 Fig. 8, page 96.

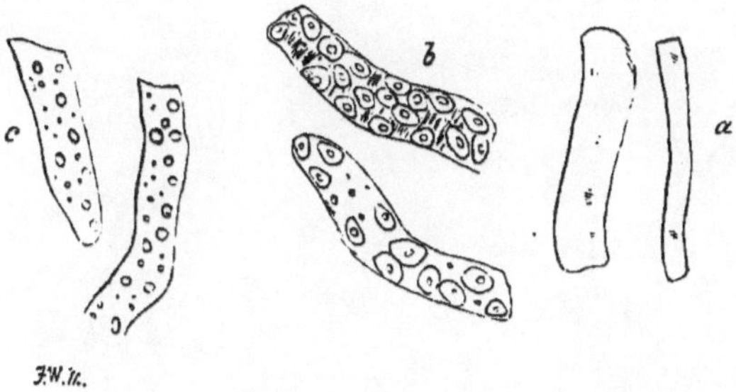

Fig. 16, page 118.

Fig. 9, page 100.

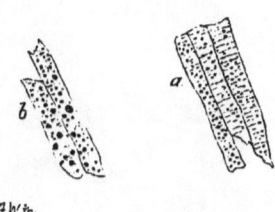

Fig. 18, page 158.

CONTENTS.

CHAPTER I.

HISTORICAL - - - - - - - PAGE 18

CHAPTER II.

PRIMAL DEPARTURES.

a. Uric Acid Formation (Reversion); *b.* Vaso-Motor Disturbance;
c. Growth of Connective Tissue - - - - 9-36

CHAPTER III.

FIRST STAGE.

THE NORSE TYPE (*Changes in the Tissues of the Mesoblast*):
Changes in the Vascular System—Joint-Gout—Rheumatism—
Chronic Bronchitis—Emphysema—Eczema—Secondary Valvular Disease in the Heart - - - - 37-57

THE NEUROTIC, OR ARAB TYPE (*Changes in the Tissues of the Hypoblast and Epiblast*): Digestive Troubles—Biliousness—
Skin Affections—Migraine—Mental Phenomena—Cardiac Neuroses - - - - - - - 58-82

Some Practical Points - - - - - - 82-88

CHAPTER IV.

MIDDLE STAGE.

a. Diseases of the Vascular System: Palpitation—Angina Pectoris —Vaso-motoria—Epistaxis—Atheroma—Aneurysm—Apoplexy— Gangrene - - - - - - - 89-102
b. Changes in the Kidney: Interstitial Nephritis—Tube-casts— The Urine—Albuminuria—Glycosuria - - - 102-133
c. Results of Toxic Blood: Uræmia—Secondary Inflammations —Gouty Asthma—Albuminuric Retinitis—Dupuytren's Contraction - - - - - - - 133-139
Some Practical Points - - - - - - 139-140

CHAPTER V.

ADVANCED STAGE.

General Considerations — Lead — Occlusion of the Coronary Vessels with Fatty Degeneration of the Heart—Arcus Senilis— The Descent—Venous Fulness—Interstitial Changes—Dropsy —Albuminuria—Serous Effusions—Death, Sudden or Slow 141-175

CHAPTER VI.

PRACTICAL CONSIDERATIONS.

a. Insurance Office View of Life ; *b.* Surgical Aspect of Vaso-Renal Change ; *c.* Relations of Stomach and Liver ; *d.* Treatment—Dietary - - - - - - 176-203

L'ENVOI - - - - - - 204-216

INDEX - - - - - - 217-219

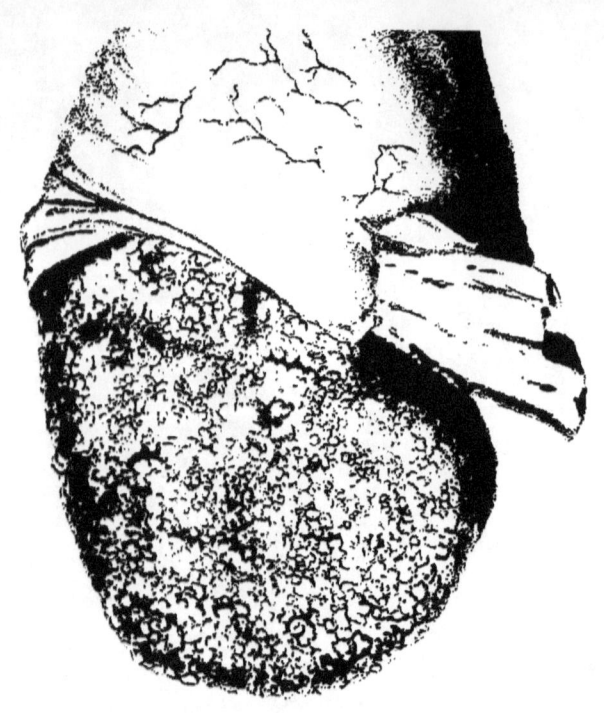

FROM THE ORIGINAL OF DR BRIGHT

VASO-RENAL CHANGE

VERSUS

'BRIGHT'S DISEASE.'

CHAPTER I.

HISTORICAL.

'So long as a disease carries a man's name it shows we know little about it,' was the remark of Sir James Paget to the writer of this monograph, when discussing its subject-matter with him, preparatory to writing it.

That many-linked chain of morbid sequences which for some time past has been known as 'Chronic Bright's Disease,' only came within the sphere of our vision in the present century. In 1813 Dr. Blackall, of Exeter, published a treatise on 'Dropsy,' in which he has a great deal to say about 'coagulable urine.' But in a great many cases the kidneys were not examined after death. In other cases they were examined and pronounced 'sound,' where it is morally certain from the histories of the cases that some interstitial nephritis must have been present, but passed unrecognised. In other cases the renal organs were observed to be diseased, as the following passage shows: 'The kidneys, likewise, have been diseased in an unusual proportion in such dissections. In no less than three instances out of the eleven here referred to—viz., three in Dr. Wells's work, and eight in mine—they were thickened or hardened, and even with a confused structure, etc., two or three containing hydatids, or vesicles.'*

In 1817, Mr. James, of the same city, pointed out how a hindrance to the blood-flow in the smaller arteries caused a fulness in the larger arteries, which in its turn led to hypertrophy of the left ventricle. But these combined observations seem to have slipped out of the medical mind of that time. And it was not until the appearance of Dr. Richard Bright's magnificent 'Medical Reports' that the medical profession realized that changes in the kidney could be a cause of death. Dr. Bright was a great pathologist, whose attention was not devoted to the kidneys solely, and when his 'Reports' appeared (1827-32), we cannot be surprised at the effect they produced on the medical profession throughout the civilized world. The association of dropsy and albuminuria with disease of the kidney was kosmos out of chaos. He first saw their relations as cause and effect, while Dr. Blackall and others merely noted what seemed to them as coincidents. His views were based upon careful observations, and his beautiful coloured plates (two of which are reproduced in this work) attracted as much attention from their pathological as from their artistic merits.

* How near Dr. Blackall was to a complete grasp of the vaso-renal change will be shown later on. Had he but had the aid of the microscope to reveal to him the condition of the small arteries, he might have anticipated our present knowledge.

Sir Robert Christison, in his article 'Granular Disease of the Kidney,' in the 'Library of Medicine,' took much the same view, and saw several maladies 'to be occasioned by poisoning of the blood with the undischarged principles of the urine.' But he, too, saw 'as through a glass darkly,' other associations of this change in the kidney, for he wrote: 'Lastly, organic disease of the liver and heart concur very frequently with granular degeneration of the kidneys. Sometimes the one, sometimes the other disease is obviously first in origin; at other times it is impossible to say which commenced first; and occasionally the three organs are affected together, and nearly to the same degree. The most common affection of the heart is hypertrophy with or without valvular obstruction.'

The microscopic appearances of such kidneys were studied by C. J. B. Williams, Dr. Geo. Johnson, and Dr. Richard Quain. The latter wrote an article on 'The Pathology of Bright's Disease of the Kidney,' in the *Lancet* of November, 1845, dividing the renal changes into: 1, simple enlarged mottled kidney; 2, truly granular or atrophied kidney; and 3, the large flabby, fatty-looking kidney.' Of the second form, the granular kidney, he writes: ' The *truly granular or atrophied kidney*, the surface of which is rough, irregular, and generally of a pale reddish colour. In this form, the filamentous tissue, contractile in its nature, as such formations always are, exceeds the quantity of the cellular or granular matter. The latter I have observed extending to the convoluted extremities of the tubes. The contractile tissue surrounding the tubes and bodies can be readily supposed to give rise to the rough or granular formation. The form resembles the hob-nailed or gin-liver.'

There seems at this period of time to have been some differences of opinion as to the nature of the changes known as 'Bright's Disease.' For in 1849 Professor Walshe delivered a clinical lecture at University College on 'Bright's Disease not essentially a Renal Disease, but essentially and primarily a Blood Disease,' in which he made the following statement : 'I have long believed for my own part that the views which localise that affection in the substance of the kidney, and regard the anatomical changes in that organ as the essence of the malady, are one-sided and contracted. I recognise in the natural history of the disease tolerably sure evidence of primary origin in the blood.'

As to the stream of investigation into changes in the kidney, gross and microscopic, it is unnecessary to follow it here. What is rather my wish here is to keep the changes in the circulating organs well in view, which have been too little regarded by many. Rokitanski observed that apoplexy proving quickly fatal (*foudroyante*) with a large blood-clot occurred in persons with large hearts.

Latham, in his famous treatise on the heart, 1845, in speaking of disease of the arteries, points to its connection 'with hypertrophy and dilatation of the left ventricle.' Continuing : 'We see that the disease of the arteries has reached a more onward stage, and made larger and more extensive deposits of cartilage and atheroma and bone, while the liver and spleen and the kidney are found enlarged and granulated; and the transparent membranes, as the pleura and peritoneum, are thickened and opaque. And it is strange if they have not a connection with the disease diffused throughout the arteries ; and it is strange, moreover, if they have not all a connection with the hypertrophy and dilatation of the heart.' Dr. Latham was a wise and far-sighted physician, in many matters much in advance of his times.

In 1855 Professor Traube, of Berlin, published some observations on the connection (Zusammenhang) of chronic heart enlargement with disease in the kidneys. From this time a connection betwixt disease of the kidneys, of the nature of interstitial nephritis, and changes in the vascular system, and especially enlargement of the left ventricle, has been generally recognised. Traube held that the obstruction to the circulation in the kidneys augmented the work of the heart, and so led to hypertrophy. (A view which, to my personal knowledge, he greatly modified—though he did not publish these later views.) While Bamberger held that the enlargement of the heart occurred before there was any distinct obstruction to the circulation through the kidney.

Then new light came in from another side. Traube found that retained urine-salts in the blood caused arteriole spasm as evidenced by an increase in arterial tension; a view taken up by Ludwig. Urine injected into the blood, it was found, produced the same effect. The commencing point of the widespread change was now being tracked down to a change in the blood. Thus the idea that Bright's Disease was primarily a blood change (as stated by Walshe) was now made clearly apparent.

Then the microscope came on the scene. In 1862 Dr. Handfield Jones described 'Fibrosis of the Arteries'; and in 1870 a notable discussion took place at the Medico-Chirurgical Society betwixt Sir Wm. Gull and

Dr. Sutton on the one part, and Dr. Johnson, of King's College, on the other part. Dr. Johnson had been demonstrating a thickening of the muscular wall of the arterioles in cases of Bright's disease with an enlarged left ventricle. Sir Wm. Gull and Dr. Sutton exhibited sections of arterioles which had undergone another form of change which they termed 'arterio-capillary fibrosis.' A brief warfare raged over the matter, and the controversy attracted general attention to the subject. Sides were taken by men who knew very little about the matter; while a few investigated the subject. Among these were Drs. Quain, Garrod and Broadbent, in England; Professors Rutherford and Grainger Stewart, in Scotland; and Dr. Ringrose Atkins in Ireland.

Clinical observations grew more acute in the meantime; and Dr. Dickenson wrote: ' Since I have had my attention particularly directed to this subject, I have scarcely seen an instance in which, if the renal state was distinctly recognised, whether after death or in life, some degree of cardiac hypertrophy was not also apparent. I have, in fact, got to regard simple cardiac hypertrophy as one of the most important diagnostic signs of renal fibrosis.' That the increased arterial tension of this state produced an accentuated aortic second sound, was pointed out by Rosenstein. The increased bulk of urine of this condition Traube showed was due to this heightened arterial tension.

In 1872, in the first edition of my work, 'The Heart and its Diseases,' is to be found a chapter devoted to the consideration of 'Combined Heart and Kidney Disease,' and descriptive of the condition to which the term 'The Gouty Heart' was applied in the second edition (1879).

Fifteen years of further attention to the subject may be a sufficient vindication for the writing of the present essay.

In the meantime the study of the sphygmograph was yielding good fruit. The late Dr. Mahomed rendered himself specially conspicuous amongst a band of workers, including Sir Walter B. Foster, M.D., Professor Burdon Sanderson, Drs. W. H. Broadbent, Galabin, Goodhart, and others. Dr. Mahomed was attached to the London Fever Hospital, and made a study of cases of scarlatinal nephritis. From this he proceeded to read a paper before the Medico-Chirurgical Society (1874) on 'The Etiology of Bright's Disease, and the pre-Albuminuric Stage,' which attracted much attention. Dr. Goodhart said in his Bradshawe lecture (1885) on 'Morbid Arterial Tension' (which lecture as originally intended should have been delivered by Dr. Mahomed himself, but his untimely death prevented it) 'it bears directly upon the present part of our subject. In it was shown precisely that morbid condition of the pulse (high tension) precedes any evidence of disease in the kidney; and he draws the conclusion from a number of observations, that the vascular condition is the cause of the albuminuria, and not the converse, as has been generally supposed.'

The present position of our knowledge on this subject is about this :—

A waste-laden condition of blood causes contraction of the tiny arterioles, producing high arterial tension. This, in turn, leads to enlargement of the left ventricle. The sustained high arterial tension sets up atheromatous change in the arterial wall; and from this spring a number of maladies. Spasm of the arterioles gives *angina pectoris vaso motoria.* The enlarged heart is liable to secondary valvular change—the result of strain upon the valves. While the blood laden with waste, or excrementitious matter, possesses distinct toxic properties, giving us gout in all its protean forms; secondary inflammations of serous membranes, bronchitis, and skin affections. Lithæmia and uræmia are part of the troubles occasioned by the blood-change. The growth of connective tissue at the expense of the other structures of the kidneys in time works their ruin, giving us intestinal nephritis, and with it what Sir Andrew Clark terms 'renal inadequacy.' While albuminuria and glycosuria are not uncommonly found. Ultimately there is heart-failure, and death.

Some more detailed account of these morbid stages will be given later on; but from what has been already stated it is abundantly clear that the term 'Chronic Bright's Disease' is no longer adequate, but is 'one-sided and contracted;' and therefore that some new term is necessitated by the further extension of our knowledge; without any intention to disparage the work of the excellent pathologist and clinical observer who first noted this morbid change.

CHAPTER II.

PRIMAL DEPARTURES.

a. Uric Acid Formation (Reversion); *b.* Vaso-Motor Disturbance; *c.* Growth of Connective Tissue.

THE view that the origin of 'Bright's Disease' was connected with the blood has steadily gained ground, until it now holds the field unquestioned. Further, it is hardly in dispute that the *materies morbi* is uric-acid.*

The Uric-Acid Formation.—This has been hitherto termed 'gout,' because the first recognition of it began with those changes which appealed to the eye. Drop by drop (goutte), something from the blood was deposited in the articular cartilages of the small joints, producing obvious deformity. The great toe was the favourite locality for this *materies morbi*, while the knuckles were also commonly affected. Occasionally deposits were found in larger joints. A like deposit was recognised on the pinnæ of the ears (otolites). Points exposed were apparently selected. These exposed points appealed to the eye, which made the diagnosis of gout, or gravel.

It was not until chemistry had cleared up the composi-

* What part, if any, is played by the earlier members of the descending series of albumen-metamorphosis ending in uric acid and urea, is unknown.

tion of this *materies morbi* that the essence of the malady was recognised. When it was ascertained that the deposited matter was urate of soda, the next step was to find out if this substance existed in the blood of these 'gouty' persons. This matter was demonstrated by Dr. Garrod. With this discovery came new light. The old idea, that 'chalk-stones' (as these formations of urate of soda were termed) were deposited drop by drop from the blood, was seen to be correct. They were the outward and visible signs of a hidden condition of the blood. Now it became possible to understand the relation of certain subjective matters to visible gouty formations. There were bodily sensations and mental phenomena, as Sydenham knew, linked with the uric acid formation, and relieved by an outbreak of articular gout. The connection, as a clinical fact, was distinctly recognised. Other portions of the body were affected injuriously by the uric acid circulating in the blood. Gout was now recognised as a blood condition; still the old term lived on tenaciously, being prefixed by the adjectives 'latent,' or 'suppressed,' or 'irregular,' or 'retrocedent.' Advancing knowledge entailed, as it ever must, a changing phraseology; and with that some inevitable confusion.

At this point it may be well to ask the pertinent question, Why are exposed joints liable to be the seat of deposits of urates? Why do the urates in the urine fall as a sediment to the bottom of the chamber utensil during a frosty night? They are not obvious when the urine is voided; they do not become visible till a certain temperature is reached. If the urine be warmed they disappear. The liability of the hands and feet; and the pinnæ of the ear to be chilled is certainly a part at least of the cause of the deposit of urates in them. But there must be something more than mere liability to a low temperature, for we do not find chalk-stones in the nose, which is equally exposed to the action of a low temperature, on the one hand; and we do find the aortic valves to be liable to gouty changes, and they surely are never chilled. There must be something in the nature of the cartilage to invite deposits of uric acid. But so far my inquiries of various histologists has failed to elicit any explanation.

Then again, stone and gravel were matters which appealed to the eye, and were at an early period recognised as associates of articular gout. They show conclusively that uric acid can crystallise at the temperature of the body.

From these considerations we can understand how the uric acid formation first revealed itself in obvious discernible morbid changes; and how it came about that these were slowly recognised to go hand in hand with certain inward changes and subjective sensations, as matters of clinical fact.

What is this uric acid formation? For long the connection of gout with high feeding, and the well-known fact that it lay hid in the shadow of wealth, obstructed a right and true conception of its real nature. It was held as something added to a healthy person, as a consequence of indulgence of the palate. The fact of gout occurring with abstinence was recognised as conflicting with this view; and as a partial solution of this anomaly the phrase 'poor man's gout' came into fashion. So long as the theory of gout being something added to a person by high feeding held sway, so long was it simply impossible to comprehend why the uric acid formation came to be found in persons of notable abstinence and small eaters. The fact that the children of gouty parents were liable to inherit gout was a small contribution towards a solution. 'The fathers have eaten sour grapes and the children's teeth are set on edge.' If the offspring of gouty parents practised great abstinence as to eating and drinking, they more or less completely escaped the clutches of their hereditary foe. Those who indulged suffered for it. Those who practised abstinence escaped. That persons of gouty descent can keep the gout at bay by poor living is as true, as that a man may acquire gout for his descendants. Take the case of the poor gentleman of straitened means who is never attacked by the gout, except when he goes on a visit to a wealthy relation, when a few days of good living evoke acute articular gout; and contrast it with the case of the plebeian-born, self-made man, who eats and drinks at will, and himself knows nothing of the gout. How comes this about? The one has an inheritance of good living, of ancestral livers well-employed for generations. The other is born of poverty and scanty fare. The one has what Dr. Budd called an 'insufficient liver;' the other has a liver in all its integrity. With the former a trifling indulgence brings out the inheritance of a tendency to uric acid formation; while the liver of the latter remains capable of meeting all the demands made upon it. The plebeian-born man acquires the gout with his fortune—and leaves them both to his descendants.

The uric acid formation then is closely linked with an incompetent liver. That is another contribution to our acquaintance with this complex subject.

If a person possesses a functionally feeble or insufficient

liver, that fact renders him or her liable to the uric acid formation without any indulgence of the palate. Many persons who eat but little possess a great tendency to gout in some form in consequence of their liver disability. Now the 'poor man's gout' stands unveiled before us, *i.e.*, the uric acid diathesis. Gout poison has its associations with 'blue blood;' and the person liable to 'poor man's gout' is often the inheritor of a naturally incapable liver as the result of gastronomic indulgence on the part of his ancestors.*

The matter of liver insufficiency will engage our attention at greater length when dealing with vaso-renal change in the 'Neurotic;' when we shall see small spare beings who eat but little, and who often instinctively loathe animal food; yet they are liable to the uric acid formation. What, then, is this uric acid formation which has two associations: (1) 'The rich man's gout,' where the liver is chronically overburdened by food excesses; and (2) 'The poor man's gout' where an insufficient liver has no great burden to bear? It is reversion, or a falling back on the part of the liver.

Dr. Lauder Brunton writes me: 'Another point of interest in regard to reversion is the tendency of children to form uric acid before, or within a few days after birth. This tendency appears to indicate a primitive uric acid formation; because the early life of the embryo appears to correspond with the life history of the race.' This is linked doubtless with the fœtal circulation being that of the higher reptiles.†

* We shall see farther on that another factor than good-living may play a causal part.

† The significance of uric acid infarcts in the renal tubules has been discussed by medico-legal writers.

In the lower forms of life excretion seems as comparatively simple as digestion; and highly differentiated organs are not found. But as the ascent of creation is made we find the alimentary canal becoming complex, and bearing on it elaborate organs; just as we see the general surface undergoing modifications. Certain areas become specialised for the various senses. Other areas possess merely general sensibility. While an involution of the surface gives us the urinary apparatus. We see indeed in the development of the embryo the outer layer—the epiblast having its medullary and epidermal sublayers; the nervous system and the sensitive skin. We see the inner layer (hypoblast) giving us the glandular organs of organic life. This tells of descent from the primitive endoderm and ectoderm of the *Gastrula*. The mesoblast, or middle layer, gives us the locomotor apparatus, the vascular system, and the genito-urinary organs.*

Thus we find a liver developing on the alimentary canal; and a portion of the general excretory surface of lowly life specialised into the urinary tract. (That the sweat and the urine should possess much in common is readily intelligible.) When these differentiated organs appear special functions can be traced.

When we see kidneys developed we find with them, as a consequence, the urine. When urine first appears it takes a solid form. Uric acid, as urates, belongs to animals with a solid urine. The cold blooded reptile and the bird of high temperature, alike, find their excrementitious tissue waste, or albumen-metamorphosis, in the

* As we shall see, later on, the uric acid formation runs along the tissue derived from the different layers of the early embryo in different individuals.

form of uric acid.* But when the mammalia appear we find a fluid urine, and the form of renal excrement the soluble urea. Still there remains a certain uric acid formation in all mammals; and a small amount of uric acid is normal in the healthiest of men. In the herbivora we find hippuric acid as a sort of intermediate form. The urea formation is higher than the uric acid formation. Urea is eminently soluble. Uric acid is infinitely less soluble; and is a form of excretion unsuited to kidneys constructed to cast out soluble urinine-salts in a fluid urine.

So long as the proportion of uric acid in solution is small the kidneys remain uninjured. But when—from any cause—the liver reverts, or falls back to the uric acid formation, then injury to the kidneys is the result. Years ago, Dr. George Johnson, F.R.S., pointed out that kidney changes were the results of the products of liver-imperfection in many instances; and this becomes quite intelligible after the foregoing considerations. These comparatively insoluble urates are cast out by the kidney, or retained in the blood. If the former, the kidneys become injured in time; if the latter, these urates are deposited in the body as gout. Very often something of both occurs; and so common is the granular kidney (Bright's Disease) found along with gout, that the term 'gouty kidney' has been applied to it. The association of the two is easily comprehended.

The liver seems, to a certain extent, to be worn out by long, hard service, and no longer capable of (practically) complete urea formation; and falls back, or reverts, to the uric acid formation—which is probably easier. What is

* A small quantity of urea begins to show itself in the bird.

acquired by the parent is often transmitted to the child; which steps into its inheritance—be the same what it may! It may be wealth and the uric acid diathesis. It may be poverty extending back for generations with a perfectly capable liver. The wealth has its drawback; the poverty its advantages!

Another question to be asked is this, 'How is it that the uric acid formation is so marked in the English people?' Gout is much more common in England than elsewhere. Pointing out one day a typically gouty man (an old brewery man) to an American lady doctor, she asked, 'How is it that we have so little gout in my country?' A Scotch medical man (himself gouty) replied for me, 'Because neither in your country nor mine have we been rich long enough to be gouty.' There was no doubt a great deal in this. But there is something more. The Anglo-Saxon has always been fond of toothsome morsels. His oxen, his sheep, his pigs, and his fowls are the best of their kind. As a breeder he was always prominent. And this was as much the outcome of his palate as his pride of ownership. The cattle and sheep of Continental Europe yield meat which will not admit of being served up as the solid joint, the pride of the English matron who boasts of her 'plain roast and boiled.' It has to be served up in 'fragments, and dressed as an *entrée*.' And these 'greasy messes' (as the English housewife contemptuously terms them) are infinitely less given to set up gout than the solid joints—the chop and steak of old England. Her geographical position, plus the enterprise of her people, led to the acquisition of wealth, which was not dissipated by the march of armies and the waste of war, as was the case with every area of the Continent. It is her prosperity, taken in connection with her flocks and herds, that has made England the land of gout *par excellence*. It is the extent of the uric acid formation which has given the English physicians the opportunities of seeing vaso-renal change of which they have availed themselves. Of course, it is seen in other countries; but not to a like extent. While the dietary of the inhabitants of warm climates consists largely of hydrocarbons, the beef-eating Englishman is the gouty man. And what part has been played by the form of ale-brewing in England, it is not possible to say precisely; but this fact is known, viz., that many people who possess 'livers' can drink with impunity beer brewed on the foreign or Lager plan, who have to abstain altogether from English ales—or take the consequences.

The first departure, then, in vaso-renal change, physiologically, is the reversion of the liver to the lowly uric acid formation.* Anatomically, the first departure is the development of connective tissue, the lowliest material of the body, at the expense of the other and higher tissues.

Vaso-motor Disturbance.—The main permanent features of chronic Bright's Disease, or vaso-renal change, are a tight artery, a large left ventricle, and the physiological outcome of high arterial tension, a large bulk of urine. These are well recognised clinical facts, as the works of Marey, Galabin, Mahomed, and others have shown, as to high arterial tension.

That the *materies morbi* (uric acid, in all probability, ac-

---

* The association of the uric acid formation with deterioration is shown by its common occurrence in children of the strumous diathesis ; a race of tissue inferiority, with an osseous system, as Laycock pointed out, of infantile or lowly ethnic form.

companied by other forms of nitrogenized waste) in the blood irritated the vaso-motor system of nerves, has been recognised from the earliest day of acquaintance with this morbid change as a whole. But so far no attempt has been made (at least to the writer's knowledge) to associate the two, and see in what relation they stand to each other. Yet it seems that the two facts of the presence of a *materies morbi* in the blood and high arterial tension stand in the most instructive relation to each other. They suggest that the first consequence may be really and truly a self-preservative depurative action on the part of the system. This is a somewhat startling matter at first sight; but the more it is looked at the clearer and more vivid does it become.

Uric acid is comparatively insoluble; but in ordinary normal amount it is dissolved in the fluid blood, and got rid of by the renal secretion without much difficulty. When it exists in abnormal quantity, a larger bulk of urine is required to get rid of it, and cast it out. And how is this brought about? Actions go on in the body which seem to give intelligence to the tissues and to certain arrangements, and in none more strikingly than in the attempt of the system to clear itself of uric acid. The increase in bulk of the muscular fibres of the heart to overcome a difficulty has been described as a sort of species of intelligence. The so-called intelligence is nothing more than the self-preservative power which the body possesses; and which has been developed through long periods of time in the survival of the fittest. Not only is it seen in the main action of the vaso-renal change, but it is palpable in many of the outcomes which have been, and were regarded as diseases *per se*.

Dr. Bence Jones described the rise of temperature in acute articular gout, and the inflammation, as converting the joints for the time being into so many supplementary kidneys, *i.e.*, that the action was a depurative action. Acute gout, secondary inflammations, uræmic vomiting, and uræmic diarrhœa, are marked instances of self-preservative action on the part of the system, when the blood is surcharged with an excess of nitrogenized waste.

The first effect of the accumulation of the products of albumen-metamorphosis is to irritate the vaso-motor centre, with the result of vaso-motor contraction, and a rise in the blood-pressure within the arteries. Or maybe it is the presence of impure blood in the vasa vasorum which directly irritates the vaso-motor nerves, with resultant contraction of the arterioles. The walls of the bloodvessels contract upon their contents; the pressure within the glomeruli of the kidney is increased, and, with that, the bulk of the urine. A larger bulk of urine carries with it a larger quantity of the insoluble uric acid. Just as when sugar is present in the blood in excess thirst is the result; and the thirst secures the imbibition of water, which washes the offending sugar away. When the liver is out of order, malproducts are present in the blood, with the resultant consequences of loss of appetite, which eases its burden; and thirst, which causes fluids to be taken in considerable quantities, and so the peccant matter is swept away. These actions no more involve intelligence than do hunger and thirst. They are illustrations of the capacity of the system within certain limits to take care of itself.

In what other way could the system increase the bulk of urine than by raising the blood-pressure in the arteries?

By no other means that we know of in the present state of our knowledge! Thirst leading to the drinking of fluids no doubt raises the blood-pressure in the arterial system until the free flow out by the kidneys once more balances matters. By its irritant or stimulant effect upon the vaso-motor centre, uric acid is cast out of the blood.

This view of the long complex process, lasting often over years and many years, is highly instructive as to the production of many maladies, which are the direct outcome of this attempt upon the part of the system to right itself; to free itself from the presence of uric acid—the primitive urine-stuff which does not rightly belong to the mammalian body with a fluid urine.

Pathology is physiology modified; and when uric acid is circulating in the blood-current, the system possesses a means of getting rid of it. But the effort entails many morbid changes, and departures from health as the outcomes thereof; of which some account will be given in the following pages.

But while recognising the broad fact that high arterial tension and a copious urine are characteristics of the vaso-renal change, there are variations therefrom, according to certain modifying circumstances, to be considered shortly. That uric acid accumulates in the blood, and is removed by an attack of acute gout, the observations of Garrod show conclusively. The uric acid seems to be broken up into urea and oxalic acid, which escape by the kidneys. How far the system possesses the power of breaking up uric acid into soluble matters, and so clearing itself, is a matter on which much darkness still rests. Why the change in arterial tension is less marked in some persons than others, and, with that, its resultant phenomena, may depend upon this breaking up of uric acid. The system may possess several means of getting rid of uric acid; and one be more operative in one person, while another takes the lead in another person of the uric acid formation. For certainly we see the change as regards the vascular system more pronounced in some persons than in others. In a complex matter involving many outcomes — which the vaso-renal change most certainly is — no one individual can manifest all the phenomena; because some are antagonistic to others, as we shall see. Still, high arterial tension is the condition *par excellence* in vaso-renal change. As the kidneys become extensively injured, the output of both the soluble urea and the insoluble uric acid falls, and the urine is of low specific gravity; while the blood is imperfectly depurated. In other words, the power of the system to secure blood-depuration is waning; just as the hypertrophy of the left ventricle, to meet the high arterial tension, wears out in time—and the heart fails in its energy. Indeed, some individuals die with marked kidney-changes; while in others, again, it is the heart which attracts the attention, and heart-failure which is the cause of death, and not the kidney.

Whatever objections may be taken to the view that the rise of arterial tension is a self-preservative action of the system as regards the later stages of vaso-renal change, there seem valid grounds for so regarding it as a primal departure, at any rate.

That diminished output, in addition to a tendency to uric acid formation, may be essential to accumulation in the blood, and excitation of the vaso-motor system, is probable. As Dr. W. H. Dickenson informs me that

'Disease of the liver, without disease of the kidney, such as cirrhosis, which is presumably associated with excess of uric acid in the blood, certainly does not cause hypertrophy of the heart or arterial thickening.' If the kidneys, then, cast out uric acid, the accumulation in the blood, with its consequences, does not exist. Sir. Wm. Roberts, M.D., thinks 'that in "high tension" the blood contains much "dirt"—of very varied character—not merely uric acid, but also much else that would be better cast out by the excretory channels.' That uric acid is accompanied by other members of the descending series of albumen-metamorphosis in the accumulation of waste matter in the blood, has been suggested already in this chapter.

Direct experiment alone can positively determine the association of excess of uric acid in the blood with high arterial tension and increased bulk of urine. I have not been able to hear of any such experimentation. But Ustimowitsch (Ludwig's 'Arbeiten' for 1870, p. 212) found that the injection of urea into the veins of a dog invariably caused a rise of blood-pressure. For this information I have to thank Dr. Lauder Brunton, F.R.S., who thinks it possible that it may be found that uric acid possesses the same property, even to a higher degree.

COMMENTS UPON THE DEPURATIVE THEORY, BY DR. F. W. MOTT.

The theory that the diuresis of chronic Bright's Disease, or vaso-renal change, is a depurative action on the part of the economy, is an attractive one. But it is a mere speculation that the uric acid accumulation in the blood stimulates the vaso-motor centre, thus raising the general blood-pressure, and causing an increased flow of urine.

The notion that the diuresis which occurs causes more of the insoluble uric acid to be washed out, forming a self-adjusting mechanism for its elimination, is, unfortunately, unsupported either by clinical experience or by physiological investigation; although there are many facts in favour of a depurative theory when applied to the soluble nitrogenous waste-products, such as urea.

Brücke, in his lectures, was the first, I believe, to suggest that the abundant diuresis met with in many cases of hydro-nephrosis was a physiological compensatory action, by which the urinary salts were washed out.

Cohnheim, by his experiments, has proved that the composition of the blood has a marked influence on the renal circulation. The injection of urea into the circulation of an animal causes decided increase in the amount of blood flowing to the kidneys, and an increased flow of urine. From these experiments he is of opinion that the degree of dilatation of the smaller renal arteries is dependent upon the amount of urinary products which the blood happens to contain; consequently, when parts of the renal cortex have undergone destruction, as in chronic Bright's Disease, there will be a dilatation of the small renal arteries of the remaining healthy portions of the organs, allowing a large quantity of blood still to circulate through them; this dilatation of the small renal vessels being induced by a constant tendency to accumulation of urea, and other nitrogenous waste-products, in the blood. The experiments to which I have already alluded support strongly the theory that a physiological compensation is effected by a dilatation of the small renal vessels.

The amount of urine secreted by the kidneys is dependent upon the pressure in the glomeruli. This pressure may undoubtedly be raised by any cause which leads to an increased flow of blood through them. Consequently those glomeruli which remain healthy will receive relatively a very large quantity of blood at a very high pressure for the following reasons: A hypertrophied left ventricle, acting with increased force, drives into the renal arteries of the small gouty kidney as much blood as the normal heart drives into the renal arteries of the healthy kidneys; the difference being that the pressure in the former is greater than in the latter.

But if the same volume of blood is forced into the renal arteries of the atrophied kidneys, as into those of the healthy kidneys, the effect will be that the pressure and the rate of flow of the blood in those glomeruli which remain healthy in the contracted kidney will be very much greater than in the glomeruli of the healthy organ. This exaggerated blood-pressure in the glomeruli of the contracted kidney will have the following effects: An increased flow of urine takes place, which will serve to wash out the nitrogenous waste-products, especially such as are soluble; but at the same time the organ will suffer from this continuous high pressure, and the renal inadequacy must of necessity be a progressive one.

This physiological compensation, however, can only be effectual when combined with the maintenance of a corresponding resistance in all other parts of the arterial system, such as we find to exist in vaso-renal change, or Bright's Disease, even in its earliest stages. If it were not for this increased peripheral resistance, induced either by obstruction to the flow of blood through the capillaries, or by arterial spasm, the effect of the dilatation of the renal arteries, brought about by the influence of the accumulated urinary salts in the blood, would be in a great measure nullified. This is because the blood would be determined away in the direction of least resistance, and consequently less blood would flow to the kidney. Really, the compensatory mechanism allows a large quantity of blood to pass through an arterial area of a very much diminished capacity, the result being a rapid flow at a high pressure through the glomeruli, and resultant polyuria.

While admitting that the augmented flow of urine which occurs in chronic Bright's Disease, due to the contracted kidney, has many arguments in its favour, I hardly think that this theory of physiological compensation can be applied in the way stated by Dr. Fothergill. Having previously suggested this theory in somewhat different words, I have considered it untenable for the following reasons :—

It is as yet unsupported by experimental investigations; or, so far as I know, by clinical experience and observation.

The following evidence appears contrary to the uric acid hypothesis. Uric acid is an extremely insoluble salt, only 1 in 15,000 parts being dissolved by water, consequently five pints of water will only take up about $3\frac{1}{2}$ grains. If the water be made acid, it would take up still less; therefore the amount of uric acid eliminated by the kidneys depends not so much upon the quantity of water which is excreted, as upon the degree of acidity, and the condition in which the uric acid exists in the blood.

As to the cause of the accumulation of uric acid in the blood in gout, there are three views:

1st. An increased formation of uric acid, owing to imperfect oxidation.

2nd. A defective excretory power of the kidney for the elimination of uric acid.

3rd. An accumulation in the blood of uric acid, owing to a deficiency in the alkalinity of the blood, rendering the uric acid more insoluble, and therefore more difficult to eliminate. This deficient alkalinity of the blood has been attributed to an increased formation of lactic and volatile fatty acids which are produced by digestive disturbances, so common in gouty people. The last view is the most plausible from clinical and experimental observations.

Sir Wm. Roberts has shown in health that the amount of uric acid excreted by the kidneys is three times as great during the alkaline tide following a meal than at other times. Again, the administration of potash and lithia salts in the treatment of gout supports this view.

If the uric acid while circulating in the blood acts in the way suggested on the vaso-motor centre, causing a rise of blood-pressure, the gouty patient should from the very first pass large quantities of water. On the contrary, lithæmia is usually associated with scanty high-coloured urine depositing an abundance of lithates, not on account of the increase in the proportion of uric acid eliminated, for that is usually diminished, but being thrown down, on the urine cooling, on account of its acidity.

As the disease progresses, the quantity of uric acid eliminated per diem diminishes, and finally in the late stages may be absent altogether. Now it is particularly with late stages of gout, when the kidneys are small and granular, owing to the cirrhotic changes in their structure, that the increased flow of urine occurs.

This, then, would be the period when little or no uric acid is eliminated. It appears to me that another means is adopted by the economy for purifying the blood of this waste-product. As soon as the uric acid has accumulated to a certain extent it combines with soda to form urate of soda; and this insoluble salt is deposited in a crystalline form from the lymph in the cartilages of the small joints. Later on in the disease, it may be detected in the substance of the kidneys between the renal tubules, thus leading to degenerative changes in the organs. After a paroxysm of gout the patient is relieved, and the blood is to a great extent purified. In the intervals of the paroxysms, uric acid is accumulating in the blood; consequently, if we accept this hypothesis, there should be a progressively increasing exciting cause acting upon the vaso-motor centre, with a corresponding increase of arterial tension and flow of urine. But does this occur?

No doubt the retained *materies morbi* does increase peripheral resistance either by the effect it produces on the walls of the capillaries, or by the stimulation of the vaso-motor nerves causing contraction of the arterioles.

How far this is due to the uric acid, and not to the renal inadequacy, cannot be determined; but this much is certain, that in cases of obstruction of the ureters and production of the obstructive form of uræmia, we have no evidence of high-tension pulse, although the *materies morbi* are accumulating.

Again, in the pre-albuminuric stage of scarlatinal nephritis, we have high-tension pulse before any sign of renal disease appears. Gout is, in most cases, accompanied, or followed, by renal disease. We cannot prove that the high-tension pulse and the enlarged left ventricle would exist were it not for the latter complication; and as all forms of renal change, except the lardaceous, are accompanied from the very earliest condition by a high-tension pulse, it will be quite impossible to attribute with any degree of certainty the increased peripheral resistance entirely to the circulation of waste-products in the blood.

In many cases of leukæmia there is a considerable increase of the uric acid eliminated by the kidneys, perhaps induced by defective oxidation processes; yet there is no increase of arterial tension or of the urine passed.

The increased flow of urine only occurs when, owing to the degenerative changes in the kidneys, there is a tendency to the accumulation of urea in the blood; and if we term Cohnheim's compensatory a 'depurative' theory to rid the system of the waste-product, more especially the soluble urea, I am of opinion that there is very much to be said in its favour; but this is totally different from the hypothesis which I have been asked by Dr. Fothergill to criticize.

These comments are valuable in demonstrating how much we have yet to learn before a satisfactory hypothesis can be broached as to the association of high arterial tension with the accumulation of products of albumen-metamorphosis in the blood. The subject is well worth investigation by clinical observers, as well as by experimental physiologists, who have found urea to raise the blood-pressure in the arteries. Whether my own hypothesis is unwarranted by facts, as Dr. Mott holds, and Cohnheim's is the correct solution; or both are insufficient, and the correct hypothesis has yet to be found; it is my firm conviction that ultimately the high arterial tension of vaso-motor change will be found to be a part of the self-protective action on the part of the organism; which, 'traced from within its inmost centre to its outmost skin,' the vaso-renal change is from beginning to end.

Growth of Connective Tissue.—In order to comprehend clearly this growth of connective tissue to the detriment of the normal and higher tissues, it becomes desirable to glance at the tissue-development of the embryo. At a very early period three layers are seen: (1) The outer or epiblast, which furnishes the central nervous system and the sensitive epidermis—in other words, the means by which the organism is in communication with its environment; (2) the inner, or hypoblast, which gives the glandular apparatus of organic life, *i.e.*, the glandular organs which lie along and debouch into the alimentary canal; while (3) the middle layer, or mesoblast, furnishes the locomotor apparatus, bone, muscle and cartilage, the vascular system, and the genito-urinary organs. From the wander-cells of the mesoblast springs the lowly connective tissue, or packing material of the body, which holds the other tissues together.

As the embryo develops, we see how from the epiblast the axis-cylinder of a nerve projects, while the mesoblast throws round it nerve-sheath and bloodvessels. From the hypoblast we can see the glandular structure of the liver gradually forming, while the mesoblast is supplying the fibrous stroma, or framework of the gland, and the bloodvessels; which not only feed the glandular structure, but which bring to it at a later day the material upon which it exercises its proper function. The different tissues are woven together, the elements of each remaining distinct. There is an interweaving; but not a blending, or fusing one into another.

The following engraving is taken from the 'Elements of Embryology,' by Foster and Balfour. It represents the development of the eye of a rabbit.

'In the invagination of the lens a thin layer of mæsoblast is carried before it, and is thus transported into the

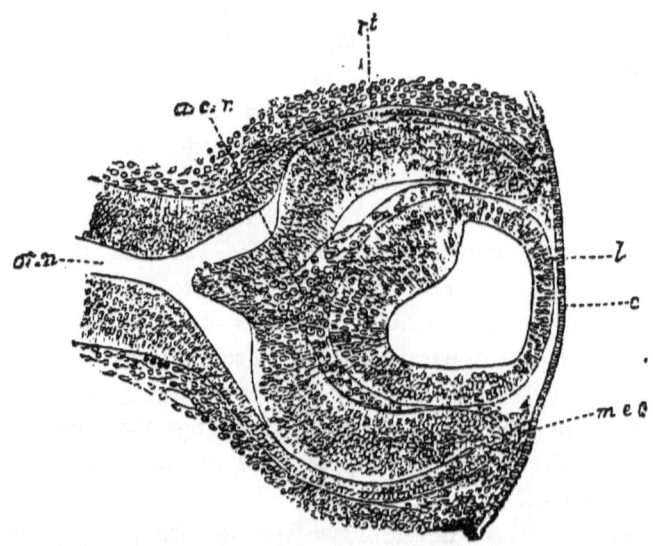

Fig. 1.—Section through the Eye of a Rabbit Embryo of about Twelve Days. *c*, Epithelium of cornea; *l*, lens; *m, c, c*, mesoblast growing in from the side to form the cornea; *rt*, retina; *a, c, r*, arteria centralis retinæ; *op, n*, optic nerve.

cavity of the vitreous humour. In the folding in of the optic vesicle, which accompanies the formation of the lens, the optic nerve is included, and on the development of the cavity of the vitreous humour an artery, running in the fold of the optic nerve, passes through the choroid slit into the cavity of the vitreous humour. The sides of the optic nerve subsequently bend over, and completely envelop this artery, which gives off branches to the retina, and becomes known as the *arteria centralis retinæ*.'

The consideration of this engraving teaches us how the different structures are built up together like a house of various materials, while the connective tissue holds all together. In the vaso-renal change we find the connective tissue encroaching upon the other tissues. It grows luxuriantly at their expense. The high blood-pressure, which is found more or less in all cases of vaso-renal changes, is accompanied, or followed, by a growth of connective tissue. We find it in the arterial walls, rendering them in time rigid, inelastic and brittle—from whence arise many of the diseases which spring from arterial degeneration (atheroma), as apoplexy and aneurysm. We see it extending through the kidney, gradually and slowly progressing; till at last the organs are so extensively injured that they can no longer do their work, and the organism perishes.

Holding the urates within it, the blood acts as an irritant upon kidneys constructed for the elimination of soluble urea.

'In studying the *causes* of Bright's Disease, we have to look for an irritant; and it is here important to observe, in the first place, that the inflammatory manifestations occur in both kidneys; and, although on examining the kidneys in detail, these manifestations will be found to present themselves in innumerable minute foci, and to confine themselves for the most part to the cortex, yet they do not affect a few definite areas, but are diffused over the length and breadth of the organ. These facts indicate that the irritant is carried to the kidneys by the blood, and is distributed by the blood. As the blood is primarily distributed to the cortex, and as the cortex contains the more active secreting tissue of the organ, it so happens that the inflammatory manifestations occur almost exclusively there.'

In the various cases of Bright's Disease we do not find the various constituents of the kidney tissue equally engaged; and it is necessary in studying the causation to refer to this matter. In the kidney we have, as the main constituents, the bloodvessels, the secreting tissue (consisting of the glomeruli and the tubules, with their epithelium), and the connective tissue. It is not impossible that an irritant brought to the kidneys by the blood may show a predilection for the renal epithelium on the one hand, or the connective tissue on the other. But it seems more likely that in the majority of cases it will act on both at the same time. We have already had several opportunities of observing that chronic inflammations very commonly produce inflammatory new formations in the connective tissue, and the kidney forms no exception in this respect. It may be said, therefore, that for the most part irritants which act through long periods, and with little intensity, produce a chronic inflammation mainly of the connective tissue. On the other hand, irritants which act intensely, so as to produce acute inflammation, while they produce the usual changes in the bloodvessels which we have seen to occur in acute inflammations, affect mainly the epithelium of the uriniferous tubules. It may therefore be said that acute inflammations are *parenchymatous*, while chronic inflammations are *interstitial*.

This distinction, however, cannot be rigidly carried out. There are some acute inflammations which affect the interstitial connective tissue even more than the secreting epithelium; and in all acute inflammations which are somewhat prolonged, the connective tissue is involved. On the other hand, inflammations which have begun acutely, frequently become chronic, or subacute; and in them, while the epithelium continues to show marked changes, the interstitial tissue always shows distinct inflammatory proliferation ('Manual of Pathology,' by Coats).

Dr. Coats here differentiates in a lucid and instructive manner betwixt the acute inflammation of the kidney, which, commencing in the tubules, involves in time the connective tissue (which might fairly enough be spoken of as 'Chronic Bright's Disease'); and the interstitial chronic inflammation due to an irritant borne upon the blood-current, which, though affecting other structures, is 'a chronic inflammation mainly of the connective tissue' (and which would be more correctly spoken of as the renal factor of vaso-renal change). Of course we can easily see how primary renal disease, by causing imperfect blood-depuration, may in some cases be the starting-point of vaso-renal change. With primary renal disease, however, we are not here concerned.

The accompanying drawing shows very clearly the formation of one of these minute foci located in the cortical portion of the kidney. The overgrowth of this lowly connective tissue would matter comparatively little if it merely grew, and added to the bulk of a viscus. But unfortunately all connective tissue of pathological origin carries with it an innate tendency to contract; as is seen from the cicatrix of a burn downwards. It is this contraction which compresses the other tissues within its remorseless clutch. A lush growth of connective tissue involves contraction; and, with that, the compression of normal tissues to the point of their destruction, as regards their functional capacity. The piece of imprisoned normal tissue is lost for ever to the rest of the viscus. It is like

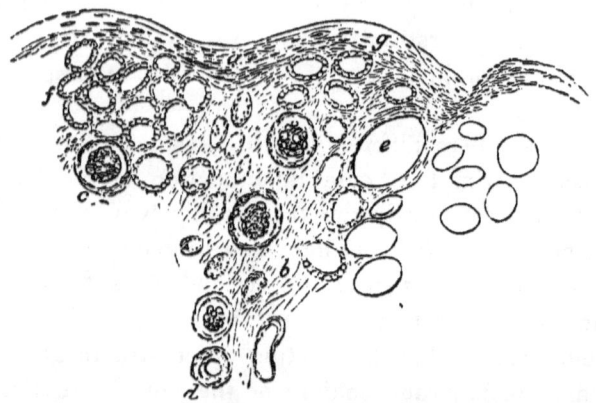

Fig. 2.—Section of the Cortex of an Advanced Granular Contracted Kidney.—The capsule, *a*, is thickened, and elevations and depressions are seen, the latter corresponding to the situation of the glomeruli, around which the fibrous tissue, *b*, is more abundant. *c*, Glomeruli; *d*, vessel with thickened walls; *e*, small cyst forming *f*, tubules, many being atrophied. Magnifications, 80 diameters.

the garrison of a beleaguered fortress, hopelessly surrounded and cut off from the retreating main army, which it can never rejoin.

In the engraving this invasion of new connective tissue is well seen at *a*. The tubules are compressed, and the glomeruli obliterated, while the contraction has dragged in the capsule of the kidney, causing that granulation of the surface which gave rise to the term 'granular kidney.'

At *b* the growth can be seen around a glomerulus not yet compressed. At *c*, the compressing process has commenced. At *d* is clearly seen the thickening of the arterial wall by a development of connective tissue within it. At *e* is a tiny cyst formed from an occluded urinary tubule. The work of destruction is going on underneath the eye. A portion of the kidney has been destroyed. The work of spoliation is progressing apace. It is merely a question of time. Sooner or later, according to the intensity of the irritant, the work of destruction will be practically complete; and the kidney injured, until it is unequal to keeping the system alive. The intensity of the irritant will be profoundly influenced by the extent to which the uric acid formation exists.

The greater and the more extensive this destruction of the kidney, the less perfect will be that depuration of the blood which is the special work of the kidney. And with this, as we shall see hereafter, those secondary inflammations of serous surfaces and elsewhere, which constitute one of the many dangers that overhang those who are the subjects of vaso-renal change,—when this morbid condition of the blood becomes pronounced and permanent.

While the growth of connective tissue is the essential anatomical departure from health, the condition of the blood starts up changes in the circulation. The impure blood excites, as we have seen, peripheral arteriole spasm; and this leads to an obstruction to the blood-flow out of the arteries. This damming of the blood in the arteries opposes an obstruction to the outflow of blood from the left ventricle on its systole; and this in turn leads to hypertrophy of the left ventricle. This condition lasts usually many years; and is compatible with great energy, mental and bodily. But the infiltration of connective tissue growth into the wall lessens its elasticity, and includes the aorta. The growth is often very marked in the coronary arteries, which become tortuous, by which the blood-current is slowed; while the reduction of the bore, or lumen, of the arteries, by growth of connective tissue in the inner coat, leads to a lessened blood-flow. Between the two the heart is imperfectly fed, and a process of fatty degeneration in the ill-nourished muscular wall is set on foot; which leads, in its turn, to failure of the heart's action, and with this, following in its wake, the final scenes of this long pathological process.

Whether it is the wall of the arteries of the heart, or the kidney,—the growth of connective tissue is the ruin of both alike in time.

CHAPTER III.

FIRST STAGE.

THE NORSE TYPE (*Changes in the Tissues of the Mesoblast*): Changes in the Vascular System—Joint-Gout—Rheumatism—Chronic Bronchitis—Emphysema—Eczema—Secondary Valvular Disease in the Heart.
THE NEUROTIC, OR ARAB TYPE (*Changes in the Tissues of the Hypoblast and Epiblast*): Digestive Troubles—Biliousness—Skin Affections—Migraine—Mental Phenomena—Cardiac Neuroses.
Some Practical Points.

SEVERAL points of interest will reveal themselves as the progress of the morbid sequences of liver-reversion is considered. Whether there are other products of faulty assimilation, or earlier members of the descending series of albumen-metamorphosis—which commences in kreatin and kreatinin, and ends in uric acid and urea—playing a part in the vaso-renal change, or not, it is impossible to say. But certainly there is a reversion to the uric acid formation, with the attempt of the system to cast it out.

Urates circulating in the blood set up a vast variety of morbid actions and processes; of which the first to be observed were such as appealed to the eye. Cheiragra and podagra, with tophi on the ears, were matters which could readily be determined as to diagnosis. Then the association of these with certain subjective feelings and moods was recognised, as Sydenham pointed out. Irritability of temper and depression of spirits were the concomitants of a fit of acute gout; while, on the other hand, he found that whenever he applied himself to literary work he brought on the gout. The Earl of Chatham was positively melancholic for the space of two years, and retired from public life, till a fit of gout restored him to his wonted mental state and enabled him to return to his place in the legislature.

Then it was observed that the vascular system was involved, when the stethoscope brought cardiac murmurs within the reach of practical medicine. The fact that eczema was linked with gouty states had long been recognised. Bronchitis, with a tendency to linger in a chronic form, was noted as a concomitant of gout. Rheumatism was frequent in men of the gouty build and diathesis.

Now it is at once curious and interesting, as well as instructive, to see how these manifestations of gout are found in the structures derived from the mesoblast, or middle layer of early fœtal life. As to the precise starting-point of bronchitis, whether in the connective tissue of the bronchial tubes, or in other structures, some question might possibly be raised; but about the other tissue-associations of gout no question can be entertained. The skeleton, the muscles, the vascular system, the dermis, the connective tissue, are all of mesoblastic origin. In the true arthritic, or gouty diathesis, the manifestations of morbid change cling to those outcomes of the mesoblast. The urinary organs have the same origin; and stone and gravel, and subsequent kidney changes, were recognised as outcomes of the gouty state. Such, indeed, was the gout of the past. It manifested a decided preference for organs of mesoblastic descent.

But there is a general tendency at the present time to move in a neurosal direction. *Tempora mutantur nos et mutamur in illis.* The massive physique of the Norseman is gradually passing away, and a lesser being of higher nervous development, but smaller in the bone, is taking its place. This is well seen in a visit to Madame Tussaud's Museum. Contrast the living crowd of small dark beings with the figures of a by-past time. These latter personages have a large skeleton, big muscles, light-coloured eyes, a florid complexion; and are just the persons to have gout in their extremities, atheroma in their arteries, and ultimately disease in their aortic valves; and die with kidney complications. As a rule, they weigh about one half more than the living beings around them. So complete is the transformation, that unless a country squire happen to be present no representative of the past is to be found in the living. In the country the old type still remains as a survival; but it is no longer to be found in town-bred people as a rule.

The direction is, at the present time, distinctly towards the smaller neurotic or Arab type. And this fact is the explanation of why the old-type gout is less commonly seen now, and new manifestations are coming into light. It is like the panoramic scenes of dissolving views under the magic lantern. One is passing away and another can be seen showing through it; at first dimly, then more distinctly, and finally in clear outline. It is easy to note on all sides this change going on. In the past rustic youths sought adventure. The Norse rover gave way to the knight-errant, the soldier of fortune; the navigator who sought distant and unknown seas. Where hard blows were to be met with, thither went the adventurers. The dangers of unexplored seas attracted others. Now we find a very different state of affairs. Men engaged in active life are seeking repose. 'The battle of modern life,' as Sir J. Crichton Browne says, 'is no longer fought with thews and sinews, but by the nervous system.' We educate our children; in the struggle for existence the brunt of the battle falls upon the brain as 'the organ of mind.' We find an enormous increase of maladies of the nervous system. It is not in the rapid growth of asylums for the insane we see the solitary evidence of this change at the present time. Indeed, we can see it outside man, in the increased nervousness of the modern race-horse. It cannot bear the rough usage—the whip and spur—which seemed to be essential of yore to supreme effort in the race. A horse which has been a few years in London will remember in fear a cuff in the face which his rustic grandfather would scarcely have heeded; and taken pretty much as the clown who inflicted it would take a cuff from his milkmaid Venus—rather as a compliment than otherwise.

Town-life has a decided effect in reducing the thews and sinews, and heightening the nervous susceptibility. And town populations now exceed rural populations in numbers. We have now a larger proportion of urban individuals to deal with; and therefore see disease in new and novel forms. Consequently, we will not be surprised to find the long series of morbid changes which spring from liver-reversion to the uric acid formation undergoing modification. It is not the products of the mesoblast which nowadays are being mainly involved. We find the nervous system derived from the epiblast to be largely implicated. Affections of the epiblastic epidermis are common, as herpes and other epidermal skin affections. The glandular elements derived from the hypoblast are involved. We get defective secretion from the gastric tubules; while the liver is liable to disturbances. These react on the nervous system; and we get bilious headaches and migraine, with neurosal disturbances of the heart. The epiblastic and hypoblastic tissues now seem to feel mostly the weight of the uric acid formation. We are seeing a new phase of this hepatic reversion. We had gout from excessive demand upon a fairly capable liver till it gave way under its burden. Now we see neurotic personages, whose congenitally insufficient livers are not equal to dealing with even ordinary amounts of food; especially animal food. There seems a development of the nervous system, with a shrinking of the viscera. In this latter case the liver reverts to the uric acid formation, just as much as in the former case.

The Anglo-Saxon in the United States of America manifests this change very markedly. He is almost free from the gouty maladies of the Norseman. But the blood of Norseman ancestors runs in his veins. He is a neurotic *par excellence*. Still lithates are as frequent (in my personal experience) in the modern American lady of slight physique, with a highly-developed nervous system, a feeble liver, and small digestive capacity, as in the burly country squire of England.

Fordyce Barker says that the people in the United States of America are not so free from gout as is often taken for granted. It is there, but it is not always recognised. Its new garb is to a certain extent a disguise. But now that that disguise is penetrated, the matter is no longer such a mystery. The uric acid formation of the functionally feeble liver can be as readily recognised in the one type as in the other, when the liver has been over-burdened. The Norseman country squire, and the petite neurosal American lady of limited assimilative power, equally manifest a tendency in the liver to descend from the urea formation to the urates of lowlier creatures.*

Clifford Allbutt, F.R.S., in his highly interesting work on 'Visceral Neuroses,' has pointed out how commonly neurotics have a family history of gout. It is not in the United States alone that this divergence from the old type to the new—the neurotic direction—is seen; it is equally seen, if not to a like extent, amidst the population of Yorkshire, in which Dr. Allbutt's practice mainly lies.

After these considerations, it may be said fairly correctly that in the gout of the Norseman, chiefly involving the tissues derived from the mesoblast, we see the gout of the past—now rapidly disappearing; while in the gout of the neurotic, mainly implicating the structures derived from the epiblast and hypoblast, we recognise the gout of the future,—now pretty clearly manifesting itself.

NORSE TYPE.

Vascular Changes.—We can now proceed to regard the morbid manifestations of the vaso-renal change (with its reversion to the uric acid formation) at some length;

---

* Dr. Lauder Brunton, F.R.S., has recalled to my mind the fact that the frog has a fluid urine and the soluble urea—a sort of shadow thrown forward by the future change.

taking the slow chronic change in the vascular system first.

However it is brought about, blood charged with uric acid does set up a certain amount of spasm in the arterioles of like character, and differing only in degree from the spasm of the *angina pectoris vasomotoria* of Eulenberg. The clinical fact that individuals undergoing this vaso-renal change are those persons who in the main manifest angina, falls in with this view, and corroborates it. Angina is a condition of acute spasm, superimposed upon a state of slighter persisting spasm.* After a certain time the result of this spasm can be seen in a distinct thickening of the muscular walls of the arterioles. That there is also thickening of the outer coat in many cases seems certain. As these coats thicken in various proportions, we may get cases of muscular hypertrophy; while in others the change might more correctly be spoken of as a fibrosis; as a very noticeable discussion before the Medico-Chirurgical Society in 1870 and 1872 has told us. Whether we are as yet in a position to speak very definitely about arterial changes, may be called in question. One thing is fairly certain,—the change does obstruct the blood-flow in the arterioles; with the consequence of a rise in the blood-pressure in the arterial system.

This change is well seen in the accompanying engraving, which is a section of a renal artery taken from a case of early cirrhotic kidney. It illustrates thickening of the muscular coat; and also of the connective tissue of the

---

* Dr. Wm. H. Broadbent thinks that sometimes an obstruction may exist in the capillaries, raising the arterial tension without arteriole spasm.

outer coat—that connective-tissue growth which is the great anatomical departure in the vaso-renal change.

From this fairy touch upon the tiny arterioles follow many and grave consequences. But at the commencement there is nothing visible. The early change is like a comet which has not yet come within the sphere of our vision. Perhaps the precise moment when it becomes visible is not always noted. But in time we recognise its existence as established. The result is a full artery; and

Fig. 3.—Section of Medium-sized Artery of the Kidney, from a Case of Interstitial Nephritis.—The middle coat, *b*, is thickened, also the tunica adventitia, *a*. Magnified 200 diameters.

with that the energy, both muscular and mental, of a free blood-supply. For years the individual goes on the type of health, very often. Others again are liable to attacks of rheumatism, or eczema, or bronchitis, or sciatica. The first stage is often a protracted one; especially in persons of the gouty diathesis, and good family history. But in others it runs its course with more rapidity, and brings its victim to the grave before the middle of life is reached. But the hearty fox-hunting country squire, who was a stout cricketer and football-player in early days, can still carry his gun over turnips, or ride to hounds, and hold his own with the best of them. The full artery means a

FIRST STAGE. 45

liberal supply of blood to all organs; and vascularity goes with functional capacity.

The full artery presents an obstacle to the blood-flow out of the left ventricle on its contraction, and this leads to hypertrophy of its muscular wall; by virtue of that law which determines hypertrophy when any hindrance exists to the ventricle emptying itself completely; *i.e.*, as completely as it normally does. (Leaving over some considerations of the changes in the left ventricle in various individuals, we will proceed at the present time with the usual sequence of events.) We find, then, a full artery with a large left ventricle; and from the high arterial tension, a forcible closure of the aortic valves and distension of the aortic root, giving us a loud aortic second sound. A tracing taken by the sphygmograph at this stage will show the square-headed tracing characteristic of the full artery.

Fig. 4.—Square-headed Tracing.

The finger feels the artery full, firm, and incompressible. The sphygmograph also tells of this,—and is the record of what the finger feels.

As a curious and instructive contribution to our knowledge of this change may be mentioned a case which fell under my own personal observation. A man of thirty-five, who had had severe unmistakable articular gout, but who was free from it at the time, was staying, ten years ago, at a large country house where the fare was liberal, the cellar well stocked, and the company good. Without any excess in food or drink, he certainly was overfed. A distinguished physiologist was one of the guests: and one day he took pulse-tracings of the company. In this particular case the pulse-tracing showed the characteristic and well-recognised square head. This was taken as a matter of course, and the fact created no surprise. But half a dozen years later, when the sphygmograph was applied to the same artery, the peculiar square head was a-wanting. The artery furnished a perfectly normal tracing. Now this observation harmonizes with the observations of Dr. Broadbent and the late Dr. Mahomed about the rise of arterial tension found after scarlatina when the blood is laden with waste matter; and which precedes the nephritis, of which, however, it is the herald. This has a most distinct bearing upon the whole question of the rise in arterial tension, and the changes which ensue therefrom when the blood is surcharged with waste matters. The fact is, indeed, very suggestive in several ways.

The condition as it stands at this stage is this—a large left ventricle; a loud aortic second sound; a full artery beginning to grow harder from the growth of connective tissue in its wall; and a free flow of urine—the physiological evidence of high arterial tension. Such a condition in a man of good family history will be maintained for years, and often many years. But the rate of progress is by no means the same in all cases. It varies very considerably from a scarcely measurable rate in some, to a course of a few years in others. In young women who pass lithates, it is comparatively rapid in its course; certainly more rapid in an urban neurotic than a rural

Norseman. A number of such cases cross my field of vision as I write.*

But it would be wrong to make the statement that pure concentric hypertrophy is the invariable change in the left ventricle. The first step is incomplete systole; and the resultant product depends (1) upon the rate at which the incomplete contraction is set up, on the one hand; and (2) the nutritive powers of the system on the other. If the demand be from a torn aortic cusp, then hypertrophy will be diluted by dilatation from the suddenness of the demand; even when the nutrition is good. Where the demand is slow, as is usual in vaso-renal change, the dilatation is arrested by hypertrophic growth before it is measurable; and pure concentric hypertrophy is the result. But this involves good nutrition; and the nutrition in a gouty man of the Norse type is usually good, and the complexion high. Where the nutrition is not good, then the hypertrophy is found with some dilatation. In women the nutrition is inferior to man; and the vaso-renal change with women is often found with imperfect hypertrophy, viz., a condition of hypertrophy blended with dilatation.

During this time there is, in the bulk of cases, little or no evidence of the change affecting the kidney; though this varies in different individuals. In some there exists a more pronounced tendency for the kidney to suffer than is the case with others. In these last, evidences of renal injury, such as albuminuria and tube-casts, or the presence of lithates, are furnished; while in many any silent change going on in the kidneys makes no sign. The urine is

* It is obvious that in all cases the progress will be slower if the uric acid formation be lessened by a proper dietary and regimen.

clear, copious, but otherwise natural. Possibly it is distinctly acid. Also, there is a tendency to get up in the small hours of the morning to empty the bladder.* Some exposure to cold may lead to some tubular nephritis; and then 'Bright's Disease' is added to the vaso-renal change, blending with it, and altering its hue and complexion; while in other cases some renal trouble may be the first departure from health. But speaking broadly, probably the bulk of cases fall under Dr. Mahomed's category of 'Chronic Bright's Disease without albuminuria,' *i.e.*, the albumen, if present at all, is only found fitfully, and in comparatively insignificant quantities.

Articular Gout.—Cheiragra and podagra were produced by deposits of urate of soda in the articular cartilages of the small bones of the extremities. Of course a certain amount of deposit must be accumulated before the change in form is obvious. Dr. Garrod, in his work on 'Gout,' shows a portion of synovial membrane from the knee, where the little specks of uric acid can be seen of the size of a small pin's head, when magnified 60 diameters. He also shows cases where, after one slight attack of gout, minute naked-eye deposits of urate of soda can be seen. Also minute deposits from the cartilage of the left great toe, which had never suffered; and the actual gout was confined to the right great toe. Though the ball of the great toe is the home *par excellence* of gout, it has other favourite habitats, as the knuckles for instance. The fore

---

° The country squire in this condition usually, when up, looks out to see what the weather is like. This may arise from agricultural interests, or he is a hunting-man, or a game-preserver; but his gamekeepers always believe he gets up to look out in order to cross-examine them about the weather; to see if they were out and about in the night doing their duty.

and middle fingers are the common seat of gouty deposits in the knuckles—the metacarpal ends of the phalanges. The proximal end of the thumb is less frequently its seat. Small nodosities may be felt along the fingers, sometimes seated in the sheaths of the tendons. The knee is most frequently affected as regards the large joints, and is also liable to synovial effusion. Both ends of the humerus may be encrusted, and so may either end of the femur. Gout in the wrist is rare, but very painful. Gouty pain is also felt on the instep and in the heel (very characteristic); and at the insertion of the tendo Achillis, and also in the aponeurosis of the gastrocnemius. Charcot has found gouty infiltration in the sheaths of nerves; and certainly sciatica is common with gouty persons of the Norse type. Dr. Garrod states 'that extensive deposits may take place within the joints without corresponding external manifestation;' and there seem good grounds for believing that in many of these cases, where pain is experienced without change of form, there is some deposit, but not enough to cause visible change of form. He continues, 'I am of opinion that not only is the deposition of urate of soda constantly found in gouty inflammation, but that it stands to it in the relation of cause rather than effect; that is to say, the deposition of the salt first takes place in the synovial membranes, cartilages, and tendinous structures of the joint, and by its presence gives rise to inflammatory action.'

Repeated attacks of gout, *i.e.*, acute gouty inflammation in the joint, tend to prevent enlargement by further deposition, to a greater or less extent. The late Bence Jones thought that such inflammation, by raising the temperature of the part, exercised an oxidizing influence over the urates; while this rise of temperature tends to render the urates soluble, and so once more to enter the blood-current, and to find their way out of the system. The inflammation is not the disease, but its consequence, 'the thunder-storm which clears the air;' and consequently Mead said, 'Gout is the cure of gout.'

One remarkable fact about gouty inflammation is this—the comparative rarity of suppuration. As a broad rule, suppuration does not occur, but occasionally abscesses do form; and the presence of pus does not positively forbid a gouty element in the case,—though it casts a doubt upon it.

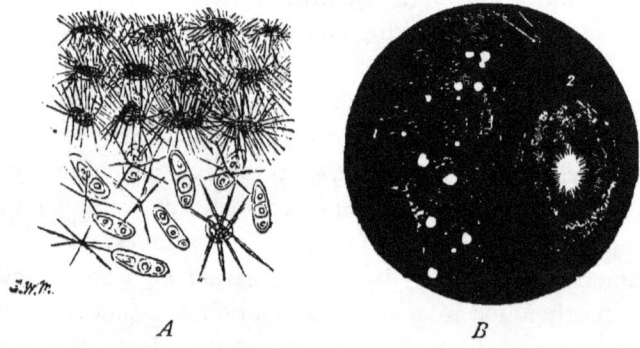

Fig. 5.—*A*. Crystals of Urate of Soda deposited in the Articular Cartilage (after Cornil and Ranvier). It will be noticed that the deposit has occurred more particularly in the cartilage cells, especially near the joint surface. Magnification, 300 diameters. *B* is the microscopic appearance of the deposited crystals within the cartilages, after Garrod. 1. 30 diameters. 2. A point of 220 diameters.

Another remarkable fact about gouty deposits on a large scale in the hands and feet, is that where these are very conspicuous the general health keeps good; and these persons usually are little plagued with gout in other forms

FIRST STAGE.

—at least until the end is approached. In one case, a foreman painter, his hands and feet were woefully deformed; but his chief complaint was of a large deposit over the middle phalanx of his right ring-finger, which was constantly getting in the way, and being knocked.

A tiny deposit on the articular cartilage, telling that the urate of soda has made its initial invasion, is shown in the accompanying drawing.

Sometimes the latent deposit gives no sign of its presence (unless it be occasional twinges of pain) until some injury to the joint is experienced, whereupon gouty inflammation sets in, and with considerable severity in some cases.

Rheumatism.—Gouty persons of the Norse type are decidedly subject to rheumatism, especially of the shoulder and upper arms. Sometimes it seems truly muscular, and confined to the muscles or their aponeuroses. At other times it has periosteal relations; and adhesions form which require to be broken down before relief from pain or freedom of movement can be attained. The arm becomes locked by them, and spontaneous movement is abolished. It is in these cases that passive movement and massage are so successful.

Lumbago is certainly frequent in such persons; and so is 'cramp.'

Exposure to cold will bring on an attack of rheumatism in persons predisposed to it; but how is it, it may be asked, that rheumatism is produced by exposure at one time, when like exposure for a score of times leads to no such result? There must be some modifying factor; and from my experience as a general practitioner in the North of England, where rheumatism is very frequent, as well as from country patients seen now, it has seemed to me that the factor is the proportion of urate in the blood. At any rate, these rheumatic persons are liable to find red lithates pretty constantly in their urine. The efficacy of the pill consisting of mercury, ipecacuan, and colchicum, given by the late Dr. Fuller in his work on 'Rheumatism and Rheumatic Gout,' is certainly another piece of evidence in favour of this hypothesis. Another is the great utility of potash in rheumatism. If anyone will take the trouble to compare Dr. Garrod's work on Gout (representing the views of the present day), with that of the late Sir Charles Scudamore (representing the views entertained half a century earlier), he will see for himself how many maladies have passed over from the class 'rheumatism' to 'gout.' My own opinion leans to the view that 'rheumatism,' of chronic or subacute character, is one of the manifestations of the uric acid formation—one of its Protean forms, indeed; whatever acute rheumatism may depend upon casually.

Many persons will admit that their trouble is 'rheumatic,' who disdain the term 'gouty' as applicable to them or their maladies. 'A rose by any other name will smell as sweet.' As a gouty man of Norse type, the writer knows from personal experience perfectly well the pain of acute articular gout, and the sudden, sharp, gouty twinge in the heel, or elsewhere; and the dull pain of muscular rheumatism. Of persistent pain he has had no experience; but its features are quite distinct, and it is easily diagnosed. That the pain of the uric acid formation should differ according to the tissue affected, is no more than reason would tell us to expect.

Chronic Bronchitis.—The relation of bronchitis, and

specially if recurrent, with the uric acid formation is now widely recognised. Sir Henry Holland observed it; and Dr. Headlam Greenlow placed the matter beyond dispute. It may alternate with articular gout, or eczema, or psoriasis. He says, 'I well recollect a striking example of this alternation in the case of an elderly man who was long under my care. His ailments were gout, psoriasis and bronchitis, and he was rarely or never free from all of them. No two of the three ailments ever co-existed in his case; but it would happen that just as he was congratulating himself on having got rid of the gout, his skin would become covered with psoriasis, and that in a few weeks would take its departure, and be succeeded by an attack of bronchitis.' Such alterations point to something causal and common to all the maladies.

'My personal acquaintance with the matter arose from my testing the acidity of the sputum as an indication for the exhibition of alkalies in the treatment of chronic bronchitis in my early days. When the scene of my labours passed from the Westmoreland hills to the Leeds Public Dispensary, I found a number of cases of chronic bronchitis—especially during the winter months—in stalwart individuals, which did not do well on the routine cough mixtures of the institution. Very frequently there was co-existent dyspepsia, or some skin affection. A suspicious family resemblance pervaded them; and in substituting a mixture containing potash for the ordinary cough mixture, the improvement, so inaugurated, strengthened the suspicion into a conviction' ('Chronic Bronchitis; its Forms and their Treatment').

Further experience has corroborated the conclusions thus formed; and the chronic bronchitic is usually a person furnishing plenteous testimony elsewhere of the existence of this vaso-renal change.

Emphysema.—Of course where there is much cough there is emphysema. After an attack of acute bronchitis in the young, a certain amount of distension of the air vesicles is constantly found. But with the elasticity of the lung-tissue, this more or less perfectly disappears; as does the emphysema of the ambitious young athletes. But as years advance the lung loses this readiness to reduce the emphysema, and each attack of bronchitis leaves the condition aggravated. The congestion of the venous system, and especially the valveless portal circulation, which is the result of impeded blood-flow in the lungs, is markedly seen in the liver; and its functions are further impaired. Not only that, but the liver is displaced downwards with the flattening diaphragm. This matter has been recognised as a clinical fact; but its recognition ought to lead to thought, inspiring a therapeutic direction. The liver occupies the warmest and cosiest nook of the body, well sheltered from cold. But when displaced downwards beyond the edge of the ribs, a portion of it is covered merely by the abdominal parietes, and so is readily chilled. 'A bilious chill' is commonly something more than a mere fashionable expression. When so displaced the exposed liver ought to be covered over carefully by a stout belt of flannel, or a cummerbund.

Eczema.—The skin of persons undergoing vaso-renal change with the uric acid formation is frequently affected. The most common and serious of these cutaneous maladies is eczema. Frequently this is the main ailment, especially when severe; and a very serious matter it may be. The lower limbs seem specially liable to form its seat. It is common in the person of Norse type, and commences in the true dermis of the mesoblast. It is an inflammation of the corium beneath the epidermis. In some persons of this class, it gives rise to the most intense itching, or to stinging pain. Some authorities say they have not remarked these two associates in skin affections in gouty persons; while others assert that gouty eczema carries with it an intensity of pain and itching.

Psoriasis palmaris is frequent in young men who are the heirs of gout. Pruritus certainly is common, as *pruritus vulvæ* in one sex, and *pruritus ani* in both. The first is probably due to the irritant properties of the urine;* while the latter is due to congestion of the portal circulation, working backward from some impediment to the blood-flow in the liver.

Secondary Valvular Disease of the Heart.—Earlier in this chapter the rise in arterial tension in vaso-renal change was described. It was seen how this usually led to hypertrophy of the left ventricle to overcome the resistance offered to the outflow of blood from the heart on systole. It was pointed out how the violent closure of the aortic valves produces a distinct modification of the aortic second sound, heightening and accentuating it; just as the pulmonary second sound becomes louder and clearer when there is obstruction offered to the blood-flow in the pulmonary circulation—whether this be due to changes in the lungs, or to mitral disease. Under such circumstances we find enlargement in the right ventricle, and loud pulmonary second sound, with atheromatous change in the pulmonary artery.

* Often containing some sugar.

The strain thus put upon the aortic valves in a number of cases leads to valvulitis. There is a premurmuric stage by which the possibility of coming valvulitis can be conjectured; and in a certain proportion of cases the tell-tale murmur puts in its appearance in a measurable time. I do not know that we can, as yet, discriminate betwixt those cases where the accentuated aortic second sound will go on and on without any evidence of valvulitis, from those where the valvulitis follows at no great interval of time. But, sure enough, in a certain proportion there is ere long evidence of valvular mischief on foot. This form constitutes the sclerosing, or progressive variety; contrasting with the static injury of primary valvulitis.* It goes from bad to worse; and in some cases but a few years elapse before the end is nigh. The strain upon the valve-curtains is such that the disease once started progresses. In three or four years the course is run in some cases; while in one case known to me, eight years have elapsed without any measurable progress having been made. We do not yet have sufficient data to determine the course of aortic valvulitis, which is secondary to and causally related to a large left ventricle in vaso-renal change. It is, however, the progressive form found with persons advanced in years. But it is not confined to persons getting on in life. After forty it is found increasing and gathering as years roll on.

In one case, well known to me, aortic valvulitis was the

---

* The two forms of valvulitis were well shown in the case of the late Lord Iddesleigh. He had a mitral valvulitis in early life, following rheumatic fever. For years, in accordance with the medical views of that time, he abstained from public life, but subsequently entered it without injury. Later on in life aortic valvulitis set in with a gouty heart; and this, in a few years, brought life to a close.

first recognised manifestation of gout. The patient, a medical man, had acid dyspepsia, flatulence, and a considerable output of lithates. Being in the habit of having a stethoscope applied to his chest, more in joke than in earnest, when in my consulting-room, one day, to my horror, I found a distinct regurgitant murmur at the aortic valves. This valvulitis was treated just as a primary attack of gout in the great toes would be. All exertion was forbidden, the arterial tension was kept down by a rigid dietary, and sweeping all waste matters out of the blood. He took an intelligent view of his own case, saw the advantages to be derived from rational treatment, and got his reward. The case has come to a standstill, at least so far as human eye can see, having been under observation four years. He leads as quiet a life as is possible to him; and permits no accumulation of nitrogenized waste in his blood.

But it is not only in the aortic valve we see this secondary inflammation from strain on the valve-curtains with a hypertrophied left ventricle. The strain upon the mitral valve-curtains can often be learned by the loud sound of their forcible closure. Sometimes insufficiency is the result; sometimes fusion of the free edges together (producing stenosis) is the result. Sometimes the case goes on measurably from bad to worse; but an extending experience tells me that this is not invariably the case. The larger the left ventricle and the greater the strain the more rapid the progress. It is in such cases that acute bronchitis is so apt to be fatal in persons not expected to succumb. When there is such stenosis with general enfeeblement of the system, the progress is decidedly slow.

THE NEUROTIC OR ARAB TYPE.

Having considered the direction taken by vaso-renal change with the uric acid formation, in the Norse type—the extreme right—it is now time to consider its career and course in the neurotic—the extreme left; recognising and not forgetting intermediate forms, blends, or hybrids. In order to more properly comprehend the increase in the neurotic class at the present day, a short historical digression is desirable. An historical digression in a pathological essay may seem to some out of place. *Homo sum, et nihil humani a me alienum puto.* A medical man ought to know something of the individual in whom disease exists, as well as be familiar with the disease. I shall, therefore, make no apology for such digression; but just leave it to vindicate itself.

When the Teutonic eruption took place we find a large massive race of strong bones and huge muscles; a hard worker, a harder fighter—fond of combat with man or beast, and of high physical courage; subduing by sheer physical prowess the slighter Celto-Iberian races of Western Europe: while his wife was the robust mother of stalwart sons. These Norsemen became the dominant, or aristocratic class of the conquered countries. The 'holy Child' of the early Spanish painters is evidently of Visigoth parentage. The 'Madonna' of the Italian painters is clearly of Teuton descent. Circumstances at that age favoured the man of stalwart physique—the work of the field and forest; the rough sports and games in which physical strength was a great advantage and gave pre-eminence. In the personal combat and the hand-to-hand fight of war the same was found; and still more the use of the mighty war-bow—to string and draw which stature was essential—favoured the stalwart Norseman; who was also fond of animal food and a hard drinker. It was in these persons, or individuals descended from them, that gout was seen and studied.

But as time went on these specially favoured persons lost many of their advantages. First came the walled town in which the slight man could live safely as a skilled handicraftsman, producing inlaid armour and weapons of war. Then came the introduction of gunpowder, which substituted the cannon and the small arm for the bow; and the big man was shorn of much of his advantage over the little man in warfare. Later still came the introduction of the steam-engine, which did away largely with the necessity for physical strength in men. The small man of acute brain is now in demand in skilled manufacture and the tending of machinery. The late John Richard Green, the historian, held that the descendants of the old Cymri were rolling back the incursion of the Anglo-Saxon who drove their ancestors into mountain fastnesses. And it would seem that he is right, and that the prophecies of Merlin are at last to be realized. The growth of education has done much to favour the small neurotic. The man who was too puny to fight in the rough old days is now in demand as a clerk and a teacher, as well as a mill hand, or a skilled artisan. Then as culture went on, we must not overlook the increasing attractiveness of the bright-brained neurotic woman.

The whole tendency of modern progress has favoured the growth of the neurotic, and decreased the advantages which of old gave the Norse type the upper hand. We see the change going on apace on all sides. The inhabitants of English-speaking America are distinctly moving in the neurotic direction; and so are the Australians. And as time goes on the neurotic will apparently kill out the Norse type; who can already be seen distinctly to be dying out as regards the 'statesmen' of the sister counties of Cumberland and Westmoreland. The old type is dying out, and a slighter race is taking their place (see p. 41). But of course it is not in rural and agricultural districts where the change of type is so pronounced. It is in our urban population that we see the change so marked and extensive. When I left the North of England for the Metropolis, I encountered a new race in the out-patient room. Instead of the stalwart fair Saxon or red-haired Dane, was to be seen a small dark race of high nervous susceptibility. The experience among the Norse type had to be tempered and modified to meet the wants of the neurotic.

With such a widespread change of type going on, we can feel no surprise at a long chronic pathological process being modified and running a different course in the Norse type and the neurotic Arab: and that the aspect assumed by it in the one is widely different in features from its aspect in the other. Hitherto, the uric acid formation has been studied in the large frame of the Norse type; where competent digestive organs have been over-taxed by indulgence of the palate, giving us the 'rich man's gout.' Now we are learning to recognise the reversion of the liver to the uric acid formation in persons of the neurotic type with congenitally small viscera, 'the poor man's gout.' It is not an accident, nor yet caprice, which is guiding the populace at the present day towards teetotalism and vegetarianism. There is something lying deeper than the will. What forms the will, and guides the choice? A profound change of type, as seen above. The beef and ale of the past are forbidden to the neurotic by a law which he instinctively obeys in blind submission.

This digression is a matter which I respectfully submit to the physicians of North America, and of Australasia. The change of type will explain—at least, so it seems to me—the modifications in food-customs found among them; the decline in alcoholic drinks; the advance in such stimulants as tea and coffee; the resort to tonic medicines, as compared to the lowering treatment of the past. It certainly has a good deal to do with the brain-overwork of which we see so much at the present time; and equally certainly it plays a large part in the diminished energies of the digestive organs; which seems—by some curious law—to go hand in hand with the high nervous development. Why the expansion of the nervous system of the epiblast should be accompanied by a shrinkage in the organs derived from the hypoblast, is not apparent. But the fact remains—*Magna est veritas.*

We must first see a thing before we can explain what it is, and how it comes about. As the neurotic becomes the great majority of the population, we can feel no surprise in the increase of nervous affections of all kinds at the present time; and that in this widespread change of type, new affections linked with the nervous system are coming within the range of our vision—certainly they are. Of that well-marked disease, 'the General Paralysis of the Insane,' with all its characteristic features, we have no account till the present century; and yet now it is a very common malady indeed in our asylums. But it does not stand alone in this respect of recent birth. Many maladies are of modern origin. Nervous disorders of the heart were first distinctly written upon by B. W. Richardson, F.R.S., who is not yet an old man; and already they form a well-recognised class of maladies.

It would seem that there is a disturbance of balance at the present day betwixt the nervous system in its demands, and the supply of pabulum furnished by the digestive organs. The systematic over-demand upon the brain leads to exhaustion, because the assimilative organs cannot supply material to the brain in a sufficient quantity; while brain-overwork certainly impairs the efficiency of the liver. What the father acquires the children inherit—in other matters than worldly goods. The hard brain-worker produces a child with a highly-developed nervous system; but with defective assimilative organs. There is the movement in the macrocosm, and the movement in the microcosm; and there seems every reason to fear that the next generation of the North Americans, especially as regards the town-populations, will manifest the neurotic type and characteristics even more markedly, and to a larger extent, than the present generation. The uric acid reversion may not always be a long-drawn inheritance from gouty ancestry stretching far away back; but may have a much shorter pedigree. It may possibly date back little more than a generation. An energetic man may build up a fortune; and with it his neurotic daughter inherits small digestive capacity, and a marked tendency to the uric acid formation—at least so it appears to me from the opportunities I have had of observing patients from the United States of America. It is not, however, confined to the daughters; there are men of the same type. Nor is this type confined to the Western Hemisphere; the neurotic is found in Yorkshire. This is Clifford Allbutt's sketch: 'The limbs are small, but often very sinewy. Such persons are as active as birds, and the absence of fat in their muscles often gives to these, in states of health, the quality of hardness under the hand. Their conversation, again, is lively, voluble, often keen and brilliant, but impressionable rather than imaginative; you may generally notice in them, too, some little blinking, twitching, or tattooing, which quickens as words and thoughts flow faster. Usually, such a patient does not readily come to you; he is brought, half-reluctant, by his wife or friends. He says, apologetically, he is an old dyspeptic, and you can do him no good. He has visited all the springs and half the doctors in Europe, and lays a bundle of old prescriptions on your desk. Once a-gate, however, his story will be a long and minute one, but never maundering, wandering or whining. His companions will tell you that he is subject to great fluctuations of the animal spirits—gay, even fascinating in society; brisk, orderly, and thorough in business, but at home, dejected or fretful. He is a small eater, a light sleeper, and a worn worker.' Every American physician will recognise the man. I will not attempt to place beside this brilliant word-portrait any sketch of the neurotic woman; who is simply the counterpart of this man with the difference of sex—a female copy, in fact, with the characteristics accentuated.

These are the neurotic or Arab beings, in whom the vaso-renal change with its uric acid formation runs a course I will now try to portray.

Digestive Troubles.—In the large majority of cases the first departure from health has been disturbance in the digestive organs. There has been a small appetite, dainty and fastidious often in childhood at an age when other children are apt to be coarse feeders. Scarcely is manhood or womanhood attained than a distinct failure, or insufficiency, in the digestive organs becomes apparent. Any attempt to entice them to take food like other people only seems to aggravate their condition. The old phrase 'to feed the disease and not the patient,' applies to this class of beings, and 'fits them like a glove.' Their feeble appetite seems the guardian of their insufficient digestive organs. Thanks to it, the latter are not perpetually overtaxed. Just as the appetite fails when the liver is upset in acute conditions, by which means the embarrassed viscus is protected, whether the condition is due to physical or to psychical causes; so in these persons a small fastidious appetite secures their livers from being overburdened. All attempts to induce them to take more food, and especially animal food (under the impression that this will make them strong), end in disaster. The neurotic has got an Arab appetite, and does best on Arab fare.* They are dosed with chalybeates in vain, and to no purpose. Vegetable tonics are apt to produce such vesical irritability that they have to be given up; as regards the neurotic women at least. Along with pain after food (as an almost constant condition, unless the strictest watch be kept against any indiscretion in diet) there is apt to be acidity and flatulence. One day the one, another day the other. I feel certain that the patient who complained to Ebstein that 'one day he was a vinegar factory, and the next day a gas retort,' was a neurotic dyspeptic, with the uric acid

* A typical specimen of the Arab type recently told me she was never so well as when she was living on the edge of a desert.

formation. So, too, in all probability was the old clergyman who used to exclaim, 'I have been a dyspeptic fifty years; thank God for it!' The cause of his pious gratitude was that he survived; while all his brothers, not so protected, had died of some form of gouty trouble. Indeed, looked at from this point of view, the indigestion is really a guardian angel, and a blessing in disguise.

All the same, it is this very digestive disturbance which sends these persons to the physician's door. Sir Thomas Watson knew this class of patient well. He wrote: 'Gouty persons are subject to various ailments, which spring from the same fountain as the well-marked paroxysm—derangements in the functions of the digestive organs, of the heart and lungs, of the brain and nerves. The most familiar of these ailments is indigestion, with its various circumstances of impaired appetite, sickness, vomiting, flatulence, heartburn, acid eructation, gastrodynia. The bowels are irregular; colicky diarrhœa being sometimes the prevailing fault, but more commonly costiveness. With all this the patient is apt to be excessively dejected and hypochondriacal, morbidly attentive to every bodily feeling, disposed to exaggerate his sufferings, and apprehensive of the worst event. When the viscera and the thorax are affected, the patient has palpitation, fits of dyspnœa, faintings, or even pangs like those of angina. In the head occur pain, giddiness, transient affections of the vision and of the hearing, threatenings of palsy and apoplexy. All these you may say are feelings and ailments to which all persons are liable. True, but the remarkable peculiarity which connects them, in some men, with gout, is this—that they often clear away and disappear upon the breaking out of a paroxysm of that disease in the foot.'

[In justice to myself and the distinguished author, I must explain that in 1871 I gave away my copy of his unparalleled ' Practice of Physic;' and it was only after this work was sketched in outline that a copy of the third edition of 1848 was purchased, and on reading over the article on ' Gout,' the remarkable fidelity of the sketch was vividly realized. Probably had I got another copy earlier, and followed it, much that has been culled from my own experience might have been made clear to me by that work.]

In this account we get in brief most of what is about to be said. The digestion, feeble at the best, is liable to be upset by any indiscretion in diet, exposure to cold, bodily exertion, or mental perturbation. As the bowels of the Hebrew of old yearned when he saw his younger brother, so the neurotic—who is a distinct approach to the Semitic Arab and Hebrew—when emotionally perturbed, feels his bowels affected.

Sometimes the pain in the stomach is experienced immediately after food, and is caused by deficient secretion of gastric juice (Leared), and is certainly relieved very largely by a dose of artificial pepsin. Not infrequently there is a certain sense of distension due to flatulence from impaired energy in the muscular wall of the stomach. At times there is distinct acidity and sour eructations, possibly due to the presence of uric acid in the gastric juice. At other times the formation of a fatty acid gives heartburn, or cardialgia, and at other times pyrosis. One American dyspeptic invariably suffered from cardialgia if he tasted warm fat; though he could eat butter when cold and enjoy it. Very often there is a good deal of hawking, with mucus, and uncomfortable feelings in the pharynx. In a certain proportion of these cases there is acidity and pain in the stomach, which is relieved by taking food. Here there is a quantity of acid mucus in the stomach. Very often nature steps in and gets rid of the acidity by vomiting; and where this is not the case an emetic gives great relief.

In other cases the pain is not experienced until an hour and a half or more after a meal, and is referred to the duodenum. Here it would seem that the contents of the stomach remain very acid; and the chyme is not neutralized by the bile but remains acid, with the result of irritation of the duodenal mucous membrane; the relief of the pain so produced being the cause why the patient seeks the physician. The complaint is also of losing flesh, from the pancreatic digestion being interfered with by the acidity of the chyme.

Very frequently the neurotic victim of the uric acid formation is troubled with chronic constipation. Very often, from the nature of the dietary, there is also costiveness. Until the bowels are regulated the indigestion and flatulence persist, no matter what care is bestowed upon the dietary. Mineral waters alone are too cold and too 'sloppy;' and a small pill, consisting of hepatic stimulants and laxatives, at bedtime, is a good preparatory step to the mineral water next morning, which should always be taken with some hot fluid.

When these neurotic uric acid dyspeptics are run down, or reduced from any reason, they are liable to severe pain along the transverse colon. The pain is often very severe; and in character seems a sort of hybrid betwixt neuralgia and colic. Sometimes it seems to be in the stomach itself, and is so sudden in its onset as to bring the sufferer to the ground. This pain seems to be experienced in the worst degree by persons who have been exposed to malaria: and one of the most marked cases that ever came under my notice was that of a very distinguished African explorer, who had had malaria over and over again.

All these disturbances are due to the presence of uric acid in the blood; and my own opinion is identical with that of Dr. Garrod on the matter. He says, ' symptoms referable to the digestive organs have already been described as being frequently premonitory of an attack of gout; but it is not uncommon to find severe and protracted dyspepsia in patients who have never suffered from a fit, but who either inherit the disease very strongly, or have sown the seeds of it by their mode of living; and in such cases the derangement of the digestive organs is doubtless frequently dependent on a gouty state of the blood. It is often a matter of considerable difficulty to make a correct diagnosis in these instances; but at times this difficulty is at once removed by the sudden supervention of a fit of articular gout, and the equally rapid disappearance of the gastric disturbance. Such relief, however, is not always experienced, and patients may suffer for months and even for years without the nature of the malady being discovered, and this occurs more especially in persons who have led a very temperate life, but who are strongly disposed to the disease: slight threatenings, however, of the toe affection are sometimes, on inquiry, admitted to have been felt, sufficient in many cases to render the pressure of the boot uncomfortable.'

Without a knowledge of the disturbances of the digestive organs linked with the reversion to the uric acid formation, a correct diagnosis in these cases is simply impossible; 'the eye can only see what it carries with it the power to see,' said the old Italian painters. And so it is in these cases. While without it satisfactory treatment is equally out of the question. And it is very desirable that there should be a more general recognition of the nature of these cases, as they are frequent,—becoming more frequent; and are, as a rule, stumbling-blocks to the profession, and especially the junior members of it.

Biliousness.—The relations of what is termed 'biliousness' to the uric acid formation are interesting and instructive. The young persons who are liable to sick headaches; to a furred tongue, a bad taste in the mouth in a morning, with sensations of depression almost amounting to melancholia; with dark offensive stools, sometimes putty-like; with high-coloured urine; of dark complexion, belonging to the class known as being of 'the biliary diathesis;' and who are liable to have spots of pain which 'can be covered by the thumb-end' betwixt the shoulders or just outside the scapula—frequently become gouty as middle age is reached. The chemical relations of the bile acids to the urates and urea link in with this clinical observation. Both the bile acids contain nitrogen and one sulphur; distinctly pointing to their descent from the nitrogenized elements of our food. That the liver which, when insufficient or incompetent, errs on the side of bile formation, should at a later period manifest a reversion to the uric acid formation, is pretty much what we might expect. That one medical man should pronounce these spots of pain—usually felt in the neighbourhood of the scapula, but less frequently over the pectoral muscle, especially on the left side, and when so occurring believed to be connected with the heart—to be due to the liver, while another attributes them to the kidneys, is intelligible enough; for they are found with products formed by the liver and cast out by the kidneys. They belong to that not yet thoroughly explored territory of the retrograde metamorphosis of nitrogenized bodies. The persons who experience these spots are either the possessors of congenitally inferior livers, or who have overtaxed competent livers; and though such persons are liable in many instances to 'bilious sick headaches,' they usually manifest a tendency to the uric acid formation, sooner or later.

On this subject Dr. Murchison, in his famous essay on 'Functional Derangements of the Liver,' writes as follows: '*Headache* is not an unfrequent result of hepatic derangement. Most commonly it takes the form of a dull heavy pain in the forehead, more rarely in the occiput, complained of as the patient awakes in the morning, and either speedily ceasing or lasting the greater portion of the day, or for several days. Such headaches are common in the subjects of lithæmia after any indiscretion in diet, or when the bowels are constipated. Their immediate cause is probably the presence in the blood of some abnormal product of albumen-metamorphosis; the derangement of the liver is usually indicated by pain and fulness in the right hypochondrium, flatulence and high-coloured urine loaded with lithates; and relief is usually afforded by mercurial and saline purgatives and alkaline diuretics.' This expression of opinion by Dr. Murchison as to the phenomena being due to some product of 'albumen-metamorphosis,' quite harmonizes with the view taken all along in this essay. This form of sick headache seems to find its proper locality in 'biliousness;' while *migraine* will be considered later on.

So much then for the troubles of the gouty neurotic as regards the tissues derived from the fœtal hypoblast.

Skin Affections.—The skin is liable to be affected in all conditions where the blood is of toxic character, and laden with waste matters containing nitrogen. The itching of jaundice is well known. So is the prurigo of elderly persons whose kidneys are probably affected. *Pruritus ani* as the consequence of hepatic derangement, is a notorious clinical fact. *Pruritus vulvæ* is also found when the urine is of irritant character, and either highly acid or charged with lithates. When there is a quantity of sugar in the urine—a very common occurrence with persons of uric acid formation—the irritation set up is often almost unbearable. Under these circumstances it is well to bathe the parts with some warm water containing some carbonate of soda, or borax, immediately after micturition. These gouty neurotics are also liable to furfuraceous formations on the epidermis.

But *Herpes* is the malady from which they mostly suffer. One case in my notebook is of special interest. It is that of a lady of thirty-two, a gouty neurotic, whose father died of paralysis, and her mother had rheumatic gout. She was a delicate child, and had glandular enlargement in the neck. At an early period she suffered from gall-stones. Her catamenia are scanty, and accompanied by much pain. She has a furred tongue usually, a bad taste in her mouth in the morning, and a foul breath. On any excitement, and she is very easily excited to a high pitch, she suffers from great vesical irritability. For the last nine years she has been liable to patches of Herpes Iris down the nerve-trunks of the left side, whenever her liver is upset more than usual. They form along the nerve-trunks of the palmar aspect of the forearm, and along the sciatic nerve. A curious matter connected with them is that they always occur upon the left side and never on the right. Her experience is that she always is best when eating little; but that she is constantly urged by friends to eat more than she feels inclined to do; and that when so induced to take more food, and especially animal food, she is always the worse for it and not the better. The fact that herpes is a malady of the epidermal layer of the epiblast (in which the terminal endings of the cutaneous nerves ramify) is a not uninteresting fact in connection with the neurotic of the uric acid formation, and liable to the vaso-renal change.

Migraine.—The matter of migraine has just been alluded to before. Sir Henry Holland in connection with migraine made this observation: 'There is reason to believe that the kidneys are the excretory organs most concerned in giving relief in these cases, and principally by an increased separation of lithic acid and its compounds.'

Trousseau regarded it as a manifestation of the gouty diathesis. Dr. Liveing, in his treatise on 'Megrim and some Allied Disorders,' regards it as linked with the condition of lithiasis. Dr. Murchison remarks on the severity of the pain; and I have seen a lady who is the subject of vaso-renal change—a gouty neurotic—go delirious with the agony of migraine. The pain in a mild attack is in and around the eye—usually the right eye. In a severe attack the pain extends to the occiput; and one migrainous lady patient put it thus: 'Whenever you feel the pain at the back of the head you know you are in for a bad attack.' Commonly flashes of light or stars are seen, especially in severe attacks; but this does not occur with all migrainous patients. Usually, but not necessarily, the attack goes off with vomiting.

There is a periodicity about migraine which smacks of lithiasis, and is analogous to recurrent attacks of articular gout. This matter has been observed by Clifford Allbutt, who has a keen eye for the neurotic in all forms. He notes some associations of migraine which quite tally with my own observations. 'A painful feature of some cases of migraine is the sense of failing inhibition which some persons feel before the explosion occurs. These sufferers are conscious for a day before the migraine of a weakened control, a horrible irritability tending to rude or passionate acts or words, which are restrained by most painful efforts. This irritability may also appear as a larval migraine, and no headache appear, so that its true nature may be undiscovered. Other persons may in like manner be at a loss to explain recurrent states of nervousness with cold extremities until they have evidence in their own persons, or in the person of a relative or friend, that such things are a fragment of a cycle which in them is happily incomplete. Irritable bladder is another symptom which accompanies migraine in some persons, and passes off with the seizure. In one of my patients this symptom occurred at times without the development of hemicrania, though clearly of the same nature as when headache followed, and it is probable that in other persons migraine might take the larval form of recurrent vesical irritation.' This last matter of vesical irritability in neurotic women of the uric acid formation is far too little known, or recognised. It is really the reason why so many ladies of this type cannot take vegetable tonics. And though they do not volunteer the statement, they will readily enough admit the fact when the question is put directly to them.

Vertigo is often experienced with migraine, but may occur independently of it. Vertigo, swimming feelings, as if stepping on a loose plank, cramp, itching, flashes of pain, or of heat, are all matters familiar to those who suffer from the uric acid formation. The pain of herpes, without eruption, is another of the plagues connected with the epidermal layer of the epiblast in such sufferers.

Mental Phenomena.—Any account of the neurotic in whom the vaso-renal change is on foot, would be incomplete without some reference to the mental phenomena. Speaking one day to a lady belonging to this class, and asking her if the psychical semeia should be omitted from the description, the prompt and decisive reply was: 'Certainly not! They play a most important part.' She spoke from a very instructive experience. When the blood is charged with bile-acids (*cholæmia*) the mood is one of sadness, gloom, and depression; the horizon seems painted in with Indian ink (melancholia). When the urates are circulating freely (*lithæmia* or *lithiasis*) the mood is rather that of irritability and explosiveness. Nitrogen compounds form the explosives. And explosions of nervous energy are common phenomena. The leopard can pounce upon the antelope, which he could never overtake by chase, because the carnivora possess the power of exploding the motor centres in a very rapid manner. The sense of energy and well-being, often called 'fitness' by the athlete, which the well-fed man enjoys, has constituted —in conjunction with the palate—a large factor in the meat-eating tendencies of the Anglo-Saxon. But when hepatic reversion is on foot, then bodies—the products of albumen-metamorphosis—possessing distinctly toxic properties are formed. Whether these bodies are formed by the impaired liver, or they are formed in the alimentary canal, and passing into the portal circulation find their way through the incapable liver into the general circulation, is a matter not yet determined. Just as the liver normally keeps the bile within the portal circulation, but permits it to escape under the circumstances which produce jaundice; so with those toxic bodies. Dr. Lauder Brunton, F.R.S., has propounded the theory of 'the liver as a porter at the gate,' standing betwixt the portal and the general circulation, and acting as a filter; and the more the subject is studied the more the idea grows upon one as one studies the clinical facts, and strives to unravel them.

Dr. Murchison speaks thus of *depression of spirits* : ' The influence of the liver upon the animal spirits has been recognised by medical writers in all ages. To the belief in the existence of such an influence may be traced the origin of such terms as *hypochondriasis* and *melancholia*. Although it is not contended that the morbid states of mind to which at the present day we apply these terms, have their origin in the liver, they are unquestionably, in many instances, accompanied and aggravated by derangements of this organ.' While depression of spirits and groundless fears are the common accompaniments of hepatic derangement. In his 'Psychological Enquiries,' Sir Benjamin Brodie speaks as follows of a man with the uric acid formation: 'Uncomfortable thoughts are presented to his mind; he becomes fretful and peevish, a trouble to himself, and, if he be not trained to exercise a moral restraint over his thoughts and actions, a trouble to everyone about him. After a while the poison, as it were, explodes; he has a severe attack of gout in his foot; he is placed upon a more prudent diet; the system is relieved of the lithic acid by which it was poisoned. Then the gout subsides; happy and cheerful thoughts succeed those by which the patient was previously tormented, and these continue until he has the opportunity of relapsing into his former habits, and thus earning a fresh attack of the disease.' Dr. Murchison classes 'Irritability of Temper' among the symptoms of hepatic derangement; and Sydenham realized the mixture of irritability and apprehensiveness which forms the mood in attacks of gout. The mental state to which the inhabitants of the United States of America apply the term 'pure cursedness,' is based upon the presence of products of albumen-metamorphosis circulating in the blood. There is often an irascible temper, a mood of savagery, indeed; as seen in the latter days of King Henry VIII., who was a fairly amiable personage in his early days. A lady who experienced these abominable mental states, or moods, used to say she 'felt as if she could fight with a feather;' and the victims of the uric acid formation are often much to be pitied, especially the neurotics. The gouty Norse type of being is often a hard thinker of great mental activity and sound judgment; but hasty in temper and passionate. But the neurotic of the uric acid formation is a person who suffers much and deserves our sympathies. If they are trying to others they are a still greater trial to themselves; and we should make as many allowances for the gouty brain as we do for the gouty foot. We do not expect an elastic step with a gouty foot; neither should we expect calm mental working with a gouty brain. Indeed, it is possible to lay much infirmity of temper, much of human frailty, to the charge of uric acid.

One lady says her feeling in a gouty mood is pretty much that of 'Just let me alone;' that trifles create more wrath than their importance, or rather insignificance, warrants. Little things irritate out of all proportion to their magnitude. The mood is irascible and explosive. And especially is this the case where sleeplessness exists. Sleepless persons with the uric acid formation are angry, combative beings, greatly given to support the negative, or exhibit unbelief. When they lay their heads upon the pillow, instead of dropping off to sleep readily and calmly, they toss and turn; or the thoughts wander off, work in a circle, and come back to their starting-point, having run a useless course; simply unprofitable thinking, and contrasting with that clear, calm thought which comes to one sometimes in the watches of the night. Or they drop off and have unpleasant dreams, often awakened by nightmare. Nor is this surprising. That when the thoughts are tinged with sadness in the day, the dreams will have the hue of melancholia, is comprehensible enough. Towards the hour of rising the sleepless being falls into a comparatively sound sleep; and on getting up feels unrested and irritable and quarrelsome. No wonder, then, the lady thought the mental phenomena 'play an important part.'

Cardiac Neuroses.—Another distinct group of troubles is found with the neurotic victim of the uric acid formation, which is linked with the circulatory organs; and especially the centre—the heart. It has been pointed out in a preceding chapter, that when there is an obstruction offered to the flow of the blood in the peripheral vessels, this leads to a damming of the blood in the arteries, and a rise in the blood-pressure therein; with the consequence of hypertrophy (more or less perfect) in the left ventricle. In time there is developed hypertrophy of the muscular fibre at both ends of the circulation, with high arterial tension. But there are times when, from some cause (probably an unusual amount of nitrogenized waste in the blood), there is set up a condition of arteriole spasm, and thus we get attacks of *angina pectoris vasomotoria;* a malady, which will be reviewed at some length in the next chapter. Sometimes there is palpitation of the heart,—as the left ventricle struggles to overcome the resistance offered by the high blood-pressure in the arteries. This is the form of palpitation which occurs without effort and without excitement. It comes on when the individual is perfectly quiet; indeed, it not unfrequently wakens its victim from sleep, and comes like a thief in the night. It occurs with individuals whose cardiac hypertrophy may be perfect; as well as when the hypertrophy is diluted with dilatation. In persons whose hearts exhibit a combination of hypertrophy and dilatation, palpitation may also be excited by effort; which may mislead the unwary practitioner. But the great matter to remember is, that the palpitation of vaso-renal change is due to vaso-motor spasm; and occurs independently of effort, or of exertion, or excitement.

At other times an intermittent pulse is the result of a condition where the blood is surcharged with nitrogenized waste. It is apt to become a formed habit as years advance, and many old persons have such a pulse going on for years; constantly and unceasingly intermittent, but free from any sinister significance. The late Dr. Archibald Billing, the Nestor of the London profession, had such an intermittent pulse for a number of years before his death. But this intermittency may occasionally be found in a person of neurotic diathesis at a comparatively early period of life.

There is, however, a much more clinically interesting disturbance of the heart's action beginning to present itself to our vision; which is not spoken of by earlier writers on the heart—because, in my belief, they did not encounter it. It is only of late years that I have met with it, or at least recognised it, to put it very cautiously. Sir Thomas Watson, in a passage quoted early in this chapter, spoke of 'faintings' in connection with the nervous phenomena of gout ('Faintings, or even pangs like angina'). Consequently, these attacks must have come more or less under his notice. They are attacks to which Professor Gairdner has given the term *Angina sine Dolore*. There is none of that acute physical agony which accompanies the distension of the aortic root. The features are those of angina during the attack. The attack commences by the extremities becoming cold; the muscular system is relaxed, the patient lies flat on the back; the heart's energy wanes, the pulse is small, feeble, and compressible. It is syncope without the loss of consciousness. The sufferer has the keenest realization of the bitterness of dissolution. In angina the agony of the pain diverts the patient's attention; but here the patient lies powerless, motionless, as if paralyzed with curara; yet with the consciousness unclouded, and every sense on the stretch, till the beating of the wings of the Angel of Death become distinctly audible. I have heard such patients comparing notes, and the situation must be one of singular horror. I have had personal experience of the pain of severe articular gout; but it cannot approach the psychical suffering of these terrible experiences. I have seen even stalwart men unnerved and shaken, till they felt like an incarnate aspen leaf after an attack of this kind; and to feel utterly demoralized after them. Such a patient has lain upon the couch in my consulting-room for two hours, all but pulseless, with a bed hot-water bottle over the heart.* Another has lain down on the floor till his hour was past. After a series of them, I have seen a medical man, not lacking in courage, looking as if haunted, with his nervous system terribly shaken—so shaken that he could scarcely exercise self-command. It must be a terrible experience, this psychical pain in these attacks. In swooning, or syncope, the victim is fortunate enough to 'go off completely,' and becomes unconscious; but these unhappy beings are less fortunate, and distinctly realize all the horror of the situation. These seizures vary; sometimes the attack goes off in a few minutes after some brandy and water is taken; at other times they last two hours or more. Their recurrence is fitful. Sometimes a number occur within a few days. In other

---

* That this was really suppressed gout is pretty fairly proved by the fact that articular gout led to an anti-gout line of treatment being adopted; with the result of relief from these attacks, except in a very shadowy form.

cases the interval consists of months. We have not yet had enough familiarity with them to speak with any sense of sureness about them. The effect of uric acid solvents point to their association with the uric acid formation.

In angina there is high arterial tension and a tight artery, followed by a free flow of pale watery urine. There is, then, either (1) this arteriole spasm extending to and including the coronary circulation, and so lowering the action of the heart; or (2) the arteriole spasm is linked with some irritation of the roots of the inhibitory fibres of the vagus, for the pulse is at once slow and feeble, and the secretion of urine not increased, certainly. Whichever of these is the true solution, there is great loss of tone in the arterial system, and the blood-supply to the brain is largely cut off—leaving the sensorium in the most miserable condition of psychical distress.

In angina we know there is a distinct element of danger of the heart coming to a standstill in diastole; especially when the muscular structure is undermined by necrobiotic changes, and fatty degeneration is afoot. We know, too, that fatal syncope may occur in debilitated conditions of the system, even without any organic change in the heart. But what amount of danger to life attaches to these attacks of vascular asthenia is as yet unknown. So far as my own personal experience goes, no disaster has occurred, as yet at least. And one case is known to me where an old lady has been subject to such attacks for twenty years. At first they caused great alarm, both to herself, and all around; but the course of time and experience have allayed these fears, and all take them pretty much as a matter of course, her son informs me—who is himself a neurotic of the uric acid formation, and the victim of migraine. If they are not fraught with danger to life, and do not threaten the patient's existence, they certainly can be counted as among the most trying ordeals to which frail and suffering humanity is liable.

So much then for the maladies implicating the hypoblast and the epiblast in the vaso-renal change.

Some Practical Points.—While the attempt to contrast the course run by this change with the uric acid formation in persons of the Norse type with its progress in the neurotic or Arab type, is made in order to throw the subject up in vivid outline; there is no design to hide the clinical fact, that there is much blending and intertwining of the phenomena in actual practice. A person of the neurotic class will manifest unmistakable articular gout, or cardiao-vascular changes; while another of unmistakably the Norse type may be liable to cardiac neuroses. One of these latter, a man of as stalwart physique as any of his ancestors as far back as the Danish viking who steered his pirate war-keel to the shores of England, is liable to the seizures of the vascular failure just described. On the other hand, one of my out-patients at the Victoria Park Hospital at the present time is a pallid slight girl of seventeen, complaining of indigestion, acidity, and flatulence, migraine and vesical irritability, with her urine constantly loaded with lithates; who already has got a tight artery, a large heart, and a loud aortic second sound. What the future history of that girl will be, is of course unknown to me; but it seems exceedingly unlikely that she will ever attain length of days. Another, a slightly-built Jewess of twenty-nine, complaining in a like manner of digestive troubles, with lithates in her urine for years past, has not only got a hard artery, a large left ventricle, a loud aortic second sound; but there are evidences of secondary valvulitis (mitral). One of my patients, a Devonshire squire of far extending lineage, gave an almost ideally perfect account of the troubles of a blend, or hybrid. In early life he was subject to bilious attacks. At the age of thirty-four he had his first twinge of gout in his great toe. So little has the gout attacked his articulations, that now at the age of sixty he walks ten hours a day. For fifteen years he has had gastric acidity and flatulence, is liable to spots on his skin, has had *psoriasis palmaris*, and now has cramps; while his urine is free from sediments—as the urine of the Norse type of man is very apt to be; contrasting with the lithates found in the urine of persons of the neurotic type. In another patient, a Kentuckian, the mixture of semeia was conspicuous. He was a perfect gouty Arab, of large head and thin flank, whose father died of apoplexy. He complained of acidity and flatulence, palpitation, cold hands and feet; while his left hand burns at times; he is liable to urticaria, and subject to the most horrible sensations at times, evidencing maladies of the epiblast and hypoblast; yet he has a large heart, hard arteries and emphysema, belonging to the tissues of the mesoblast. In another case, a lady who consulted me for indigestion, acidity, and flatulence, with red sediments in her urine habitually, was seamed and scarred with eczema.

It is unnecessary to multiply cases of these hybrids, manifesting at one and the same time the troubles of both types. They may occur in any and varying proportions. All the same, this vaso-renal change with reversion to the uric acid formation runs a distinctly definite and different course in the person of the Norse type, and in another of the neurotic, or Arab type. To see a chronic and prolonged pathological process running a distinctly varying course in different classes of persons, is a clinical matter of the highest importance at the present time. Since the introduction of instruments of precision and the development of physical examination, medical teaching has gone away from the old lines of temperaments and diatheses, and attached itself tenaciously to physical signs.

The student of to-day is taught to look too exclusively to the disease; and too little at the individual in whom the disease exists. Yet the consideration of this chapter ought to induce him to ponder and reflect upon the different course taken by a protracted morbid change, according to the individual in whom it is on foot.

There is no attempt made here to ride a hobby to death in the division made of the Norse and neurotic types. The division has the approval of some of our very best observers and clinical teachers.

Another matter is this: In this consideration of the vaso-renal change the renal element has not been paraded, but rather kept in the background. This is in accordance with clinical facts. In the bulk of cases the kidney plays a passive part until the later stages are reached. Certainly in some cases the kidneys force themselves upon the attention at a comparatively early period; but this is not the rule. Why they should do so in one case more than another, may be due at times to some accidental chill or exposure, which engrafts tubular nephritis upon a system entered upon this vaso-renal change; and such seems to me were the cases Dr. Bright described in his first report. It is, of course, simply impossible to say why in one person the cardiao-vascular changes are the more pronounced; in another, articular gout is the prominent matter; while a third has chronic bronchitis; and a fourth eczema as the plague of their lives: while in other cases again, like that recorded by Dr. Headlam Greenlow, one form of trouble alternates with another in regular sequence.

One matter is curious, interesting, and, what is more, of high diagnostic value. It is connected with the epidermal section of the epiblast, viz., the constancy with which the teeth and nails are affected. The teeth are worn down

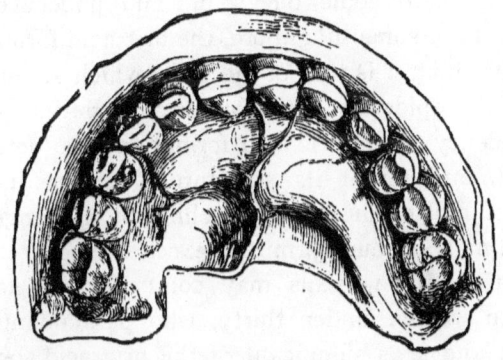

Fig. 6.

and blunted at the edge; at first slightly, subsequently till they look more like pegs than edged cutting-instruments. This change is seen in this illustration very clearly, though at an early stage. It is one of a number of casts taken by Dr. William Stewart, and exhibited by him at the Harveian Society of London, after some remarks made by myself as to the changes in the teeth of gouty persons. Persons of the uric acid formation exhibit retraction of the gum from the crown of the tooth, with a frequency common enough to give the phenomenon a distinct diagnostic value. As soon as this matter catches the eye, the clue is often given to the patient's malady; which otherwise may be obscure. This is associated with a process by which the tooth becomes loose in its socket; very often there is inflammatory osteitis at its root, and maybe a bony growth along the fang. Such teeth are liable to fall out sound, and free from caries; and in some individuals all the teeth come out in this manner. In others some come out without caries; while others alongside them evince signs of decay. But speaking broadly, when the teeth come out sound and undecayed they suggest, if not something more, the uric acid formation.

Another matter is that of the nails, which are modifications of the epidermis, and consist of agglutinated hairs. The tendency of the uric acid formation is to deteriorate the nail, and reveal its structure; whether it is the imperfect nail of the strumous child, or a change to be noted in nails of much primitive excellence. A reediness, or striation of the nails may commonly be noted in dyspeptic women under thirty, who pass quantities of lithates; and is as significant of the uric acid formation as is a gradual change in the nails of a person of Norse type fifteen years later in life.

The accompanying illustration shows this reediness, and also furnishes a very instructive story. A friend and patient of mine, Dr. John Archibald, consulted me for gouty phenomena of a mixed order in the autumn of 1884. He was put upon an anti-gout regimen, with much benefit. Soon he noticed a change in his nails. The old nail was reedy, the new-coming nail was smooth; the change dating from the time he came under treatment. When the nail was half grown—the distal end being reedy, while the proximal end was smooth—the thumb was photographed; and presented the appearance seen in *b*. A little later the reedy portion has been worn off, and the nail was smooth, *c*. As no such

a *b* *c*

Fig. 7.

modification of the nails had been anticipated, or dreamt of, of course no photograph existed of the original state of the nail. A lady of like age, and of gouty proclivities, supplied a nail bearing a very close resemblance to the original nail, *a*.*

Another sign has just had my attention drawn to it by an old physician, viz., vascularity of the inner surface of the lower eyelid. So far as my observations have gone this semeion has a value.

When there exist (1) reedy nails; (2) blunt teeth; (3) a loud aortic second sound; and (4) a distinct mental irritability at times, contrasting with the usual mood: the diagnosis of the uric acid formation (even with a clear urine) in a person of Norse type is pretty certain without any articular gout. While acidity, flatulence, indigestion and the presence of urates in the urine tell as plainly of it in the neurotic, or Arab. Such then is the vaso-renal change in its early stages.

* The artist has slightly exaggerated the striation in *b*; but certainly not as to *a*.

One more point to be attended to is the difference in the renal secretion in the two classes. The Norse squire, with his large heart and tense artery, passes a large quantity of urine, rarely depositing lithates, except at the time of a cold; whilst in the neurotic the urine usually manifests lithates.

The amount of urea passed in the first case is considerable; while less is known about its proportion in the latter instance. The Norseman, whose liver is over-taxed chronically, retains for long a good urea formation; though the tendency to uric acid formation deepens with time. In the neurotic, or Arab type, there exists an impairment of the glandular derivatives of the hypoblast from the first; and the urea formation seems defective from a comparatively early period. Consequently, though the dietary is a spare one, and animal food is eschewed—often apparently from an instinctive choice—there exist lithates habitually in the urine; and in those persons, as a very competent physician (Dr. William H. Broadbent) observed one day, 'it is almost impossible to prevent the formation of lithates.'

By an appropriate dietary and regimen the uric acid formation can be kept well in hand in persons of the Norse type; but with the other class the most careful attention to the dietary does not yield quite satisfactory results; while the uric acid solvents (potash or lithia), usually so successful in gouty cases of the Norse type, are not well borne (especially potash) by individuals of the neurotic, or Arab type. Consequently, we may fairly safely conclude that the physician can render greater services to the gouty Norseman than is possible in the case of the neurotic lady with the uric acid formation.

CHAPTER IV.

MIDDLE STAGE.

a. Diseases of the Vascular System: Palpitation—Angina Pectoris Vaso-motoria — Epistaxis — Atheroma — Aneurysm — Apoplexy — Gangrene.
b. Changes in the Kidney: Interstitial Nephritis—Tube-casts—The Urine—Albuminuria—Glycosuria.
c. Results of Toxic Blood: Uræmia — Secondary Inflammations — Gouty Asthma—Albuminuric Retinitis—Dupuytren's Contraction.
Some Practical Points.

THE vaso-renal change pursues its way steadily along the lines laid down in the preceding chapter, for years, and usually many years; my strong impression being that it runs a slower course (perhaps from its being more amenable to treatment) in persons of the Norse type, than in those of the neurotic, or Arab type. In other words, from the point of view of the Insurance Office, the Norse type is the better life. As age advances, the neurotics seem to exhibit a tendency to take on the changes in the vascular system; and to add the dangers accruing therefrom to the troubles with which they are already familiar. Consequently, a single line of progress will now be adopted as being most in harmony with the actual clinical facts.

We find (1) diseases connected with the vascular system; (2) changes in the kidney; and (3) diseases due to the vitiated condition of the blood, coming into view. And in this order will the troubles of the middle stage be discussed.

Diseases of the Vascular System.—These can be subdivided into those due to spasm of the arterioles; and those due to the waxing rigidity of the arterial walls, in connection with the large left ventricle.

The first section embraces rather disorders than diseases from a pathological point of view; but 'diseases' they certainly are, in the truest and original sense of the word, from a clinical point of view. They include palpitation, and *Angina pectoris vaso-motoria.*

Palpitation.—When the condition of arteriole spasm is set up in the hypertrophied arterioles, and the blood-flow out of the arterial system is obstructed, there is a rise in the blood-pressure within the arteries—of a depurative nature on the whole. The obstruction so offered to the contraction of the left ventricle on systole is increased; and then the heart palpitates, as the larger ventricle struggles against the resistance offered to its contraction. Such palpitation is obviously not connected with general efforts but is due to conditions of the blood; by which the arterioles are incited to contract. It is not the palpitation of a feeble muscular wall requiring digitalis, and allied cardiac tonics; but rather remedial agents which will sweep the offending matters out of the blood. Of course, where the toxic condition of blood co-exists with a feeble heart-wall, a complex pathological state may necessitate an equally complex treatment; and the two classes of remedial agents, cardiac tonics and blood-purifiers, must be used together.

Angina Pectoris Vaso-motoria.—The great matter, however, connected with the hypertrophy of the muscular wall of the arterioles, is that dread disease 'breast-pang,' or *Angina pectoris vaso-motoria*. There is a form of neuralgia of the cardiac plexus which causes acute suffering; and simulates true breast-pang, and is known as spurious angina—which lies outside the present inquiry.* Here we are dealing with *Angina pectoris vaso-motoria* (Eulenberg).

From the consideration of the effect of blood laden with nitrogenized waste upon the vaso-motor centre and the consequent rise in arterial tension, as a means of depurating the blood, in Chapter II., it would seem that *Angina pectoris vaso-motoria* is but an acute exacerbation of the chronic condition. During it the condition of high arterial tension is intensified; and a large output of urine is the resultant consequence of the attack.

Whether this may ultimately turn out to be the true explanation of the effect of blood laden with nitrogenized waste upon the vaso-motor nerves, or nerve-centre, or not, we know clinically that it is in the persons undergoing the vaso-renal changes we find true *Angina pectoris vaso-motoria*. Being a neurosal affection, it stands to reason—and experience demonstrates it as a fact—that angina is much more frequent with the neurotic than the Norse type. The famous John Hunter, the anatomist, is a case in point. Though a neurotic, he underwent the vascular changes of the uric acid formation. My personal experience of practice runs strongly in the direction of angina belonging, by preference, to the

° In my experience 'spurious angina,' or cardiac neuralgia, is essentially a malady of women; whereas true angina (vaso-motoria) is rather a malady of men.

neurotic, or Arab beings. It shows itself in them not uncommonly about the age of forty. Such sufferers are often all the more alarmed from the fact that very frequently a relative (commonly a father) died of angina. And when the sufferer arrives at the age at which his father died, *i.e.*, from sixty-five to eighty, he in turn may also die of angina. But angina in comparatively young persons whose heart-walls are structurally sound, is not fraught with much danger to life. The case of Dr. Arnold, of Rugby, may be cited as conflicting with the statement. But Arnold was a highly strung neurotic, and greatly exhausted at the time (as well as being morbidly afraid of the prospect of angina); which largely explains the fatal result. John Hunter often fainted from agony in his attacks; but twenty years elapsed betwixt his first attack and the fatal seizure. Indeed, Hunter's was a typical instance of true angina in a neurotic individual.

Such persons manifest local arterial spasms in the form of 'dead hands;' the fingers often going pallid and bloodless, and feeling, as well as looking, like the fingers of a corpse. At other times the feet go deadly cold. And one case has come under my notice where the phenomena seemed to point to spasmodic contraction of the posterior cerebral arterial area. In women, at or about the change of life, 'hot flushes' (as they are termed) suffuse the neck and face. There is acute arteriole dilatation seen passing like a wave over the visible portions of the person.* The vaso-motor system is in a state of irritability evidently in such persons.

* Whether the dilatation of the cutaneous arterioles is confined to the neck and head, and is only seen there as visible parts, is not deter-

If arteriole spasm be excited chronically by the accumulation of the products of albumen-metamorphosis in the blood, it is not difficult to conceive that times of exacerbation should occur at intervals. In all probability the immediate excitement of the spasm in the hypertrophied muscular wall of the arterioles, is a temporary excess of nitrogenized waste in the blood. The countenance is blanched, while large drops of sweat roll down the brow; the face is expressive of agony; the sufferer is incapable of movement; while the eye mutely begs for help. In a case recently in Victoria Park Hospital, the man looked like one led out for execution, and realizing the bitterness of death itself. (This was a very peculiar case apparently of traumatic origin, and not specially linked with the vaso-renal change. Otherwise, it was a genuine case of *Angina pectoris vaso-motoria*.) The radial artery is felt tense and full; and when the attack passes off, the sufferer voids a large bulk of pale urine of low specific gravity—the characteristic urine of high arterial tension (Traube).

The pain starts about the mid-sternum, and radiates therefrom. Eulenberg thinks it follows the superior cardiac nerves and the four upper cervical, also the inferior cardiac nerve and the lower cervical nerves existing in the brachial plexus. The pain is felt like a bolt through the chest, and has a tendency to shoot down the left ulnar nerve to the little finger. In one case it also went out at the ears backwards. Its duration varies from seconds to minutes; and the attack can now be cut short by such remedial agents as nitrite of amyl and

---

mined yet. Dieffenbach, when removing a tumour from the pudenda of a lady, saw a blush (Schamröthe) suffuse the genitalia.

nitro-glycerine, which dilate the peripheral arterioles; a line first taken by Dr. Brunton. In fatal cases the left ventricle is found flaccid and distended with blood; the heart stopping in diastole, or abortive systole. In one case, where the patient all but died, a condition of extreme dilatation was left behind.

The neurotic is liable to angina which may disappear under appropriate treatment; or proceed, gathering in intensity as years roll on, as was the case with John Hunter; who, not knowing its causation, could not adopt any rational preventive treatment. *Angina pectoris vaso-motoria* is an exaggerated spasm of the thickened arterioles, mostly found in the neurotic group of lithæmic persons; and comparatively little known to the Norse type until an advanced stage is reached. When the heart-wall is undermined by fatty degeneration, the danger to life is very grave.

Epistaxis.—With the condition of high arterial tension, hæmorrhage is ever probable. Bleeding at the nose in early years is a matter of trifling moment usually, and often a distinct relief to the child. But when middle age is reached, epistaxis may be serious in itself, and require energetic measures for its arrest. Beyond that there lies its prognostic import. In my own experience it has been associated with the vaso-renal change; and further with cases moving on from bad to worse at a comparatively rapid rate. An old physician of great experience informs me that it has not unfrequently been the termination of gouty persons of his acquaintance.

Atheroma.—Degeneration of the arteries commences in a growth of connective tissue which lessens the elasticity of the arterial wall. The term we use for such change in the arterial wall is *atheroma*. It occurs locally, as at the axilla and popliteal spaces, where the artery is frequently flexed; it occurs at the outside curve of the arch of the aorta, which receives the shock of the ventricular systole; and also is set up when an artery is exposed to pressure—thus the coronary arteries exposed without intermediary arteries to the full force of each cardiac systole, are often affected more seriously than other arteries. But as a widespread change, atheroma is produced by hyper-distension (Ueberspannung) of the arterial wall. It is, indeed, the outcome of the high arterial tension set up and maintained by the hypertrophy of the muscular wall of the arterioles and the large left ventricle. The effect of high internal pressure is to start up a growth of connective tissue in the wall of the artery which lessens its elasticity, and in time renders it brittle. So altered, the artery becomes liable to rupture, and this gives us aneurysm and apoplexy; two diseases which are certainly associated closely with the vaso-renal change. It is the growth of connective tissue in the coronary arteries which, at a later stage, works the structural ruin of the large heart. As pointed out in Chapter II., the first physiological departure is the reversion to the uric acid formation; while the great anatomical departure is the growth of connective tissue—which in the viscera gives us cirrhosis, and in the arterial wall atheroma.

What part syphilis plays in inclining the arterial wall to take on the atheromatous change, is a matter on which, in the present state of our knowledge, we can only speculate. Extending observation may before long bring its influence to light; but, so far, it remains largely hidden. One can even already, however, feel something more than a suspicion that syphilis exercises a malign influence in predisposing the arteries to the atheromatous change.

The accompanying woodcut shows the changes in the arterial wall wrought by a blending of syphilis with the vaso-renal change. By comparing this with Fig. 3, p. 44, the difference will at once be seen. In Fig. 3 the outer and middle coats are implicated; while in this section, in

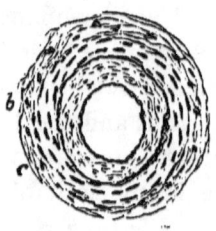

Fig. 8.—Transverse section of the Small Arteries of the Kidney. Magnification, 300 diameters. *a.* Section of an artery from a normal healthy human kidney. *b.* Section of an artery with all three coats thickened, and the lumen diminished. The specimen was obtained from a case of vaso-renal change. It will be noted that the inner coat is particularly affected, constituting a condition approaching *Endarteritis obliterans*. Probably this thickening of the interior coat was syphilitic in origin.

addition, there can be seen changes in the inner coat. It is a small vessel, and tells us much as to how the vessels of the coronary circulation undergo the atheromatous change; and especially that thickening of the tunica intima which works the ruin of the heart-wall. Beyond the growth of the connective tissue—the essential change in this morbid process—we find hypertrophy of the muscular fibre of the vascular system. And in the end, the growth of the connective tissue ruins its muscular ally.

Aneurysm.—When the arterial wall is brittle, and the internal pressure upon it abnormally high, it is liable to rupture, partial or complete. A blow at the time of arterial distension may cause this; and a fall in the hunting-field is not unfrequently its immediate antecedent. At other times the weakened wall yields, and gives us either the fusiform or the sacculated aneurysm, without rupture of any of the arterial coats. Effort certainly is productive of aneurysm; and often causes an aneurysm in the atheromatous ascending aorta. A typical instance of this occurred to a colour-sergeant who was long under my care at the West London Hospital some years ago. He was an active man of over forty, with a large heart and hardening arteries. Being called out to take part in some autumn manœuvres, he made vigorous efforts to keep up with his younger comrades. The consequence was an aneurysm of the ascending aorta. At other times an atheromatous patch undergoes fatty degeneration; and, the softened material being washed off in the blood-current, an ulcer is left, and an aneurysm forms. At other times the ulcer heals, as shown by the cicatrices formed—the 'scars of Dittrich.' The blood may find its way into the tissues of the artery, and plough its way through in time. Such a dissecting aneurysm, commencing in an ulcer in the arterial wall about two inches up the aorta, and burrowing away for some distance, came under my notice in a pallid woman of middle age, who died suddenly, when I was connected with the Leeds Public Dispensary. She had complained of vague pains in her chest, and had been confined to her bed with illness. She was improving, and was trying to dress herself, when the effort produced the final rupture of the aneurysm, with effusion into the pericardium. In this case the blood burrowed back towards the heart.

Effort, by temporarily raising the blood-pressure, is distinctly connected with aneurysms. But in my personal experience aneurysms have mainly occurred in comparatively young men who have had syphilis, and in older men in connection with vaso-renal change. In connection with the last, it has not been uncommon to find aortic dilatation, with insufficiency of the aortic valves. When we remember the distending force of the huge ventricle which goes with this valvular lesion, giving at once greater power and a larger bulk of blood, it is not difficult to comprehend how dilatation of the aortic arch is produced. In one such case, where the arterial change (general dilatation of the aortic arch, with bulging at the roots of the carotid arteries on both sides) was of recent origin, my efforts were bent towards lowering the blood-pressure by poor fare, and sulphate of soda to sweep all waste matter out of the blood, in order to see what the elasticity of the arterial wall could accomplish. Great success was thus attained, and the bulging all but disappeared. At this point the patient's wife interfered. She did not deny the general improvement—the disappearance of the cough (from pressure on the recurrent laryngeal nerve), or the diminution of the bulging in the neck. But she had made up her mind to feed him up; and left my house fully determined to take her own way, despite sundry and divers warnings of the risks she was incurring in so doing. In a month or two the patient died with a huge aneurysm under the left clavicle. She succeeded in bringing about the very danger which had been pointed out to her, and against which she had been warned. The history of the case is instructive, with both a positive and a negative lesson in it.

Apoplexy.—Apoplexy—true, genuine rupture of an intercranial artery, as compared to embolic closure of a vessel, or softening, or other cause of obliteration of the functions of the brain more or less complete—is found especially in persons in whom the vaso-renal change is established. It occurs equally alike in the small, slight neurotic and the bulky Norseman; who is supposed to be its victim *par excellence*. In the deadhouse it is quite common to see the internal carotid arteries looking like quills. In such cases there is also a large left ventricle, with some—more or less—change in the kidneys. Often the person has died of aneurysm. The atheromatous change goes on very markedly in the arteries within the skull, whose blood-current is swift. Such apoplexy is common when a sudden spell of cold sets in: the contraction of the vessels of the cutaneous area raises the blood-pressure in the arteries elsewhere, and then a brittle artery of the brain cracks. This addition to the internal pressure is the last straw which breaks the camel's back. That the blood-pressure is raised is evidenced by the increased bulk of urine experienced at such incomes of cold. The obituary column of the *Times* tells of the increase in the number of aged persons who die in sudden cold-snaps. Of course, they do not all die of apoplexy; some die of angina, or aneurysm, or sink from heart-failure, especially where the heart-wall is weakened by fatty degeneration. Still, a distinct proportion do die of cerebral apoplexy.

Very frequently small miliary aneurysms of the cerebral arteries form the seat of rupture. In the following illustration miliary aneurysms of this kind are distinctly seen.

When such alterations of the arterial walls are in existence, any rise in the blood-pressure generally will tend to rupture the weakened wall. Dr. Hilton Fagge has given a good account of such miliary aneurysms in his 'Practice of Physic;' and pointed out how they are linked with chronic renal disease. Apoplexy from rupture of the sac of these small aneurysms is readily produced by anything putting strain upon the bloodvessels—effort, straining at stool, vomiting, or arterial spasm (*angina pectoris vaso-motoria*).

Fig. 9.—Miliary Aneurysms, natural size, obtained from a case of apoplexy, by washing away the brain substance from the vessels of the basal ganglia. The patient had a large heart and contracted kidneys.

In ordinary apoplexy a blood-clot forms around the orifice. The artery keeps pumping away into the clot, acting like the hydraulic ram; and the waxing clot squeezes the soft brain-tissues, and extinguishes their function. Commonly, the motor-centres along the fissure of Rolando are, from their position and the seat of the rupture, the first to be pressed upon. Sometimes the clot is so small that these areas alone are compressed till their function is lost. As soon as the clot ceases to grow, and then becomes less by absorption, the diminishing pressure permits of the compressed portion of brain regaining its function more or less completely. At other times the clot waxes, as the blood, driven into the carotids by a large left ventricle, swells the volume of the clot. The skull is unyielding, and the imprisoned textures suffer from the presence of the intruder. This case is like that of a boat filled with passengers, when the crew want to join them, as in a shipwreck, for instance. They either throw an equal number of the previous occupants overboard, or sink the boat—which can only hold so many. So the blood-clot waxes, ploughing its way through the soft tissues, and compressing where it does not destroy. At last some interference in the respiration tells that the pressure is extending to the medulla oblongata, and that the danger to life has become imminent. The high internal pressure within the arteries is telling with fatal effect. The large ventricle is driving on the blood, and enlarging the blood-clot. The larger the blood-clot, the greater the pressure on the soft cerebral structures. The sensory area has had its life squeezed out of it; the motor area is paralyzed. The intruder ploughs towards the occiput and the base of the brain, the tentorium cerebelli giving it direction. Soon the fell pressure is experienced by the *nœud vital;* and when the medulla is sufficiently compressed, life closes.

No wonder that Rokitanski found the apoplexy which is suddenly fatal (*foudroyante*) to be linked with a large left ventricle. It is just the large left ventricle, together with the hypertrophied muscular wall of the arterioles, giving high arterial tension, which keeps up the growth of the clot, and prevents the waxing pressure from arresting the hæmorrhage. We can readily understand why apoplexy is a common cause of death in persons undergoing vaso-renal change.

Epistaxis.—Another matter linked with the changes in the vascular system, and associated with high arterial tension, is that of epistaxis. Epistaxis in persons advanced in years is a phenomenon of bad prognostic import; and one which one learns by experience to dislike. In the cases which have come under my own notice, it has occurred in persons undergoing the vaso-renal change; and has been the harbinger of disaster. In one case it was accompanied by such extensive mischief in the retina, that total loss of vision was the consequence. One medical man of large experience informs me that bleeding to death from the nose has been the actual end of a number of gouty persons of his acquaintance.

Gangrene.—The change which goes on in the arterial wall has a high interest for the surgeon. Sometimes lime salts are deposited in the connective tissue of atheromatous arteries, which renders them more brittle still; so that if there arises any occasion for applying a ligature to them, they break down, and cause a great deal of trouble. In some cases the arteries can be felt like the stems of long clay tobacco pipes. In other instances the atheromatous arteries undergo a fatty degeneration, and are thick and comparatively soft. The different form assumed by the atheromatous artery can often be readily detected in the temporal arteries meandering under the skin. Sometimes along with general atheromatous change there is a marked change at certain points, termed by surgeons 'obstructive arteritis,' which occludes the vessel; and causes the parts beyond to die, giving gangrene, or mortification.

Changes in the Kidney.—The essential change in the kidney is that of a chronic and usually slow progressive growth of connective tissue; which destroys the true renal structures. In order to make the morbid change perfectly clear to readers who may not have had recent opportunities for studying pathology, and who have been engaged in active practice, Dr. Mott has furnished to me some illustrations, which will greatly facilitate a good comprehension of what exists normally; as well as illustrative sections of kidneys the seat of this interstitial nephritis.*

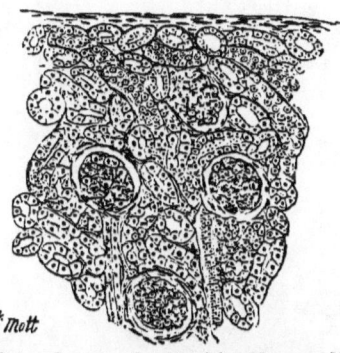

Fig. 10.—Section of the Cortex of a Healthy Human Kidney, obtained from a boy killed by an accident.—The glomeruli are seen surrounded by the uriniferous tubules. It will be noticed that the capsule is thin, and the amount of connective tissue between the tubules is so small as to be inappreciable; thus presenting a striking contrast to the sections of kidney indicating the various stages of interstitial nephritis. Magnification, 150 diameters.

In Fig. 10 we see a section of the cortex of the kidney, and see there bloodvessels and urinary tubules with the glomeruli all massed together; with a minimum of packing material, or connective tissue holding all together. Then in the following diagrammatic sketch is shown the functional apparatus of the kidney.

* They are drawn from specimens prepared by himself.

Fig. 11.—Diagrammatic Representation of a Uriniferous Tubule, and the Vascular Supply of the Cortex of the Kidney.—a, afferent arteriole going to glomerulus, and given off from b, the interlobular branch, which runs straight up into the cortex from the arch c. The vasa-recta, d, are also given off from these arches, which are situated at the base of the pyramids. The efferent vessel, e, instead of opening into a larger vessel, breaks up into a plexus around the uriniferous tubule, particularly the convoluted portion. The

Fig. 12.—Section of the Cortex of a Healthy Human Kidney, showing a Glomerulus, with Afferent Arteriole given off from an Interlobular Artery.—The convoluted tuft in which the afferent arteriole terminates, is covered by epithelium, of which the nuclei are distinctly seen. The delicate capsule is seen surrounding the vascular tuft. Around the glomerulus are the uriniferous tubules, mostly from the convoluted portion. It will be noticed how little connective tissue there is around the glomerulus and its afferent vessel. Magnification, 300 diameters.

A small artery is seen giving off one branch to the cortex, and another towards the pyramidal or tubular portion of the kidney. The cortical twig gives off still smaller twigs ending in a tuft.

uriniferous tubule, f, is seen to take a very winding and circuitous course before opening by the collecting tubule into the excretory duct, g. The portion of the uriniferous tubule, which is marked by fine dots is lined by a striated epithelium, and represents the part engaged in the secretion or elimination of the urine; while it is believed the water with some of the salts filtered out of the glomerulus, a, into the capsule, washes out the urine as it courses down the tube. Finally the solution of urine and salts, as indicated by the arrows, passes into the collecting tube, g. The line which separates the cortex from the medullary portion may be roughly considered as existing at the position of the vascular arch, e; therefore it will be seen that the cortex is concerned with all the essential functions of the urinary secretion.

In Fig. 11 this tuft is still more highly magnified. The tuft occupies a cavity, and over it and the wall of the cavity a layer of epithelium is reflected, much after the manner of the pericordium. Beyond this tuft is a larger plexus of vessels, and the obstruction offered by this venous network causes a high blood-pressure in the glomerulus itself, and exudation of fluid—the urine. The urinary tubule is coiled up among the venous network, from which it emerges and pursues a tortuous course, doubling on itself as seen in the diagrammatic sketch. Various theories have been put forward as to what goes on in these long convoluted conduits. Whether the urine is highly aqueous at the start, and collects salts of various kinds, and products of albumen-metamorphosis in its course, or *vice versâ*, is a matter which scarcely concerns us here. But by the time it has reached the straight conduits by which it debouches into the pelvis of the kidney it has become the fluid with which we are familiar.

The urine varies considerably as to its specific gravity, its colour, and its reaction even in health. In the vaso-renal change these modifications are even more pronounced; and one day it is a dilute urine, pale in colour, of low specific gravity, and copious. Another day it is high-coloured, dense, scanty, and of high specific gravity. Bearing these facts in mind, it is obvious that any examination of the urine must extend over a considerable period of time to be of any utility. The bulk passed in each twenty-four hours must be collected; and a sample be taken from the whole. The specimen of one day must be contrasted with that of several other days, with intervals between, to give any correct, or fairly accurate conception of the urine. The dietary, the conditions of life, of exercise, of quietude or effort, must be calculated with each specimen; in order to arrive at any conclusion worth the having, or the holding.

When the comparatively insoluble uric acid is constantly passing through renal structures constructed to excrete the highly soluble urea, it irritates the tissues; and a growth of connective tissue is started, here and there, throughout the kidneys.

When Dr. Bright first drew attention to the changes in the kidneys in the disease which has since borne his name, he caught sight of the final changes of the morbid process which is being sketched out in these pages. Chronic Bright's Disease was then an affection of the kidneys, and as such is classed amongst diseases of the kidney in the 'Nomenclature of Disease,' published by the Royal College of Physicians of London;* and followed accordingly by writers of treatises and text-books. Yet Professor Geo. Johnson, F.R.S., long years ago pointed and emphasized the matter in italics: '*Renal degeneration is a consequence of the long-continued elimination of products of faulty digestion through the kidneys.*' Such an affection surely cannot be a primary kidney-change; but some derangement in the organs of assimilation and of albumen-metamorphosis! Garrod says of gout and the changes in the urinary organs: 'My own observations lead me to think that gouty inflammation is often set up in the interior structure of the kidney, accompanied with deposits, not merely within the tubuli uriniferi, but in the fibrous tissue itself. It may be that the structure, from the circumstances in which it is placed, being in constant contact with a fluid

* And is so classed to the present day.

having an acid reaction, is selected as the early seat of gouty deposition; that it is frequently so chosen, proof has been afforded in the fact that white points of urate of soda were observed, with few exceptions, when these organs are examined.' Surely, in the name of reason the uric acid formation must have preceded the deposition of urates; whether in the kidney, or in the articulations?

So long, of course, as the morbid change designated 'Chronic Bright's Disease' is officially classified among 'Diseases of the Kidney,' so long must it have a misleading influence; and obscure the actual clinical facts.

That the constant presence of uric acid in considerable quantity in organs constructed to cast out the soluble urea, will excite tissue-changes in them in time, is not a matter very difficult to comprehend.

What is the tissue-change which goes on in the kidney? It is seen in a subsequent engraving, p. 112. This is what Dr. Dickenson says: 'The change begins in disproportioned growth, somewhat of a rank luxuriance in the fibroid tissue with which the arterial channels are uniformly fringed. Hypertrophy of this part leads to atrophy of all the rest. Creeping along the arterial lines with slow and hesitating steps, involving the organ not all at once, but little by little, the disease makes itself felt not suddenly, but with so gradual departure from health, that its date is usually undeterminable, and its existence unsuspected until it has reached an advanced stage.' No statement could be more explicit than this. There is a stage anterior to obvious injury to the kidney. A certain destruction of kidney-tissues goes on before any actual incompetence in the organ is reached. We all have some 'spare' kidney as we have 'spare' lung. It is only when the limit compatible with functional capacity is passed, that the kidney mischief forces itself upon the attention. How many years may elapse before the essential amount of kidney is involved may not be affirmed. So long only as the 'spare' kidney alone is being destroyed, the kidney phenomena are latent. As to the rate of progress and how it is brought about, Sir Wm. Roberts, M.D., F.R.S., delivers himself thus : ' The disease usually pursues an interrupted course. It is subject to exacerbations from time to time, with intervals of quiescence. The exacerbations are generally occasioned by exposure to cold, or some imprudence in diet or regimen; sometimes no cause can be assigned for their occurrence.' Again : 'After each exacerbation it is commonly pretty evident that the disease has taken a step in advance, and assumed a fuller development ; and that probably an additional portion of the kidney hitherto spared, or only slightly affected, has been disabled.' Virchow thinks that acute gout may manifest itself in one form as acute nephritis.

Thus we see that it is not an acute invasion (as occurs in some cases which can fairly enough be still termed ' Bright's Disease '); but a silent, insidious, stealthy process of gradual ruin of the kidney. Certainly, in such a condition of chronic interstitial nephritis comparatively slighter causes will occasion acute nephritis than are required with persons whose kidneys are perfectly sound. But it is rather a difference of degree than anything essential. Roberts, speaking of such cases, writes : ' The disease may lie concealed for an undetermined period, and then reveal itself after exposure to cold, or a fit of intoxication in the guise of an acute attack—with rapid general anasarca, and scanty sanguineous urine.' Such, to the best of my belief—so far as the history of the causes can be traced in Dr. Bright's notes—were the cases originally recorded by him in his 'Medical Reports.' They seem to have been cases of slow insidious interstitial nephritis with acute mischief engrafted thereupon. Of course when an observation is first made, the prominent phenomena are those which immediately precede death. And that excellent observer, when testing the albuminous urine in the primitive manner of a spoon and a candle, and putting two and two together in the association of the dropsy with the albuminuria, and diagnosing kidney mischief, could scarcely have dreamed of the long vista of anterior history to which he was then opening the portal. He stood at the threshold, and after him one observer after another has pushed in at the opening he made, until the unknown country has been explored; and indeed is now being to some extent surveyed. Further and further advances have been made, until now we seem within a measurable distance of its confines; and see in the reversion of the liver to the formation of the primitive urinary stuffs the first departure from health. Not only that, but we are beginning to know that excessive indulgence in animal food is not the sole cause of the uric acid formation; and that it may be the consequence of a congenitally feeble liver; and, further than that, we are beginning to have some glimmerings as to how far mental causes play a part in this liver-degeneracy. It is a far cry to Lochawe! And an immense morbid area lies betwixt Dr. Bright standing on the portal with his spoonful of urine and a candle peering into the unknown land; and the present observer already noting that vaso-renal change may fairly be regarded to a large extent as an outcome of modern civilization, and the demands now made upon the nervous system.

One observer after another has pushed the inquiry back from its finish to its starting-point—or what, at least to present vision, seems its commencement. Where it will ultimately be found to begin, however, may scarcely be affirmed positively; but in excessive demand upon the brain, and the subtle associations existing between the sensorium as the organ of mind and the glandular apparatus of organic life, one can honestly believe, if nothing more, that one can already see dimly, and as through a glass darkly, the first beginnings.

Commencing with different minute foci, tiny patches of

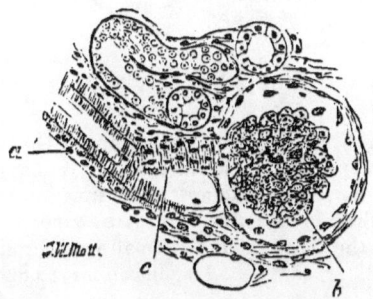

Fig. 13.—Section of Cortex of Kidney, showing Early Stage of Interstitial Nephritis: commencing Fibrous Overgrowth around the Capsule of the Glomerulus.—*a*, small interlobular artery giving off afferent branch to the glomerulus; there is, apparently, an increase of the muscular fibres of the vessel. The vascular tuft is shrunken, and there are numerous leucocytes obscuring the epithelial covering, and also lying within the capsule; and around the capsule and small artery there is an increase of fibrous tissue. Magnification, 350 diameters.

soft connective tissue, mainly in the cortical portion of the kidney, each holding glomerulus, bloodvessel and tubule in its remorseless grip,* as it contracts and hardens, it makes the part as functionally useless as if it were a piece of gristle. This is clearly seen in the Fig. 2, which is repeated here. While in its contraction the new growth has dragged down the surface of the kidney cortex, and glueing it by adhesive inflammation to the

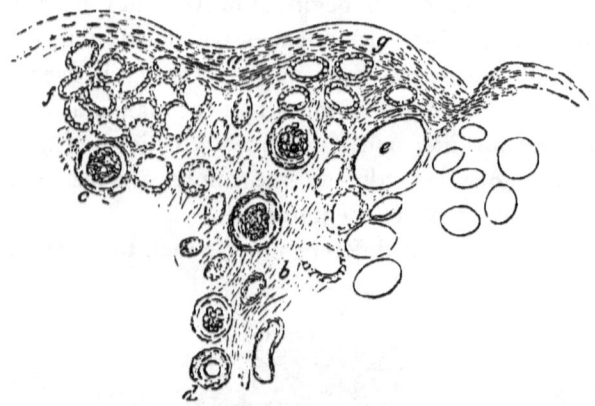

Fig. 14.—Section of the Cortex of an Advanced Granular Contracted Kidney.—The capsule, *a*, is thickened, and elevations and depressions are seen, the latter corresponding to the situation of the glomeruli, around which the fibrous tissue, *b*, is more abundant. *c*, Glomeruli; *d*, vessel with thickened walls; *e*, small cyst forming; *f*, tubules, some atrophied. *g*, capsule. Magnifications, 80 diameters.

capsule, brought that down with it; producing an uneven surface and an adherent capsule, from whence the term 'granular kidney' took its origin. In Fig. 15 we see a more advanced condition, a little deeper down in the structure of the kidney. We see a patch of the overgrowth of connective tissue holding an atrophied tubule in its clutch. We see a glomerulus destroyed as to all func-

* If the reader will compare this with the Fig. 12 at p. 105 he will see the difference, and see the glomerulus already shrinking.

tional life. We see an artery with its thickened wall, and at the top a tubule choked with colloid matter; while

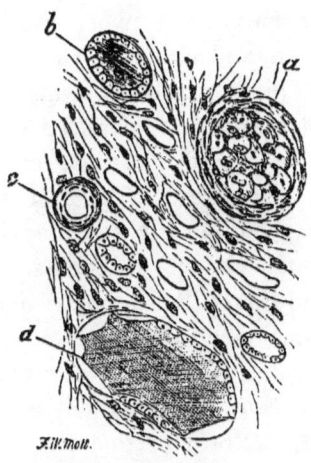

Fig. 15.—Section of the Cortex of a Granular Contracted Kidney in an Advanced Stage of Interstitial Fibrous Overgrowth. Magnification, 250 diameters.—*a* represents a glomerulus considerably contracted, and with a thickened fibrous capsule ; *b*, a tube filled with a colloidal substance ; *c*, an artery with its walls thickened ; *d*, a uriniferous tubule undergoing conversion into a small cyst. The remainder of the section is occupied by a nucleated fibrous tissue surrounding atrophied tubules.

at the bottom we see a cyst formed from a portion of occluded tubule. The engraving gives a very clear impression of the process of cirrhosis, which gradually and stealthily spreads throughout the organ. Drs. Da Costa and Longsreth have shown that this cirrhosis involves the nervous ganglia of the kidney like the rest of the organ. It is an invader which establishes itself amidst the tissue-aborigines at their expense,—like the Tartar Turk on the Balkan peninsula; and with precisely the same disastrous results, viz., extermination of the residents.

So long as enough of sound kidney remains for blood-depuration life is maintained; when the limit is reached then comes uræmia, and either coma or convulsions; or some secondary inflammation putting an end to life. Sometimes, no doubt, in less advanced cases, the inflammation is survived; and possibly, or even probably, exercises a depurative effect. But, as the limit of kidney destruction is reached, life is threatened from other outcomes of vaso-renal change, as angina, or apoplexy; for there can exist no doubt that apoplexy and aneurysm are both at times occasioned by vaso-motor spasm started by imperfectly depurated blood. Or dropsy supervenes and wears the patient out. Or maybe it is diarrhœa which closes the scene. (Not unfrequently, however, uræmic diarrhœa is the salvation of the case.) Or, in a smaller number of cases, vomiting of intractable character—the vomited fluid readily undergoing ammoniacal decomposition—is the immediate cause of death.

But such a widespread morbid change gives an almost endless variety of manifestations. In some cases the kidneys soon draw attention to them; while in other cases the patients live to the end—burn down to their sockets—without the kidney ever attracting notice; and the changes in it all along are quite latent. The kidney is not the disease *par excellence;* but only an important portion of it. Latent or obvious, the kidney plays a momentous part, as imperfect blood-depuration is the provoking cause of evil all along the line, from beginning to end. It is possible to regard the kidney as often a sort of martyr; which perishes itself in saving the rest of the tissues. The only objection to such a simile is that intelligence on the part of the tissues has figured too largely in the inquiry into the vaso-renal change. It lurked in Johnson's 'stop-cock' theory of arterial spasm; and in compensatory hypertrophy of the heart to overcome obstruction. Nature, no doubt, has conservative and preservative actions in the organism (which is self-preservative to a very considerable extent)—albeit a losing fight. It is impossible to escape from a haunting impression that the rise in arterial tension with large bulk of urine has some connection with the insolubility of uric acid.

As to the appearance presented by granular kidneys in consequence of the interstitial growth of connective tissue in them, it is well seen in the accompanying drawing, taken from Dr. Bright's beautiful plates. It is a typical kidney; and, as it appears in his book, presents a most faithful appearance, with its capsule half torn off.

The rough, uneven granular surface comes out well; while the twigs of bloodvessels on the thickened capsule are very obvious. Frequently the kidney is much less in bulk than the one here given; and especially when greater length of days is attained. The case from which the kidney was taken was a middle-aged woman, who was cut off by intercurrent dropsy.*

Facing the title-page we see a section of the other kidney from the same case (Stewart). We can see in it

---

* Oddly enough these kidneys were taken from a case where the thorax was not opened; but Dr. Bright explains that they closely resembled the kidneys of his next case, where the heart was found enlarged. The diminution in bulk goes hand in hand with years.

naked-eye changes brought about by the interstitial overgrowth of connective tissue, of which we saw a microscopical section in Fig. 14. The cortex is dragged down to the pyramids, which are themselves invaded. The morbid change is most pronounced where the two atrophied pyramids are drawn towards each other. The plate gives one a very vivid impression of the partial and unequal invasion of the kidney structures. The portions on each side of the funnel-shaped ureter were the seat of recent inflammation, or what is known among pathologists as 'cloudy swelling.' When pursuing my studies at the Pathological Institute of Vienna, the combination of recent and old mischief in the kidneys was seen recurring in a very large proportion of cases.

The microscopic appearance of the destroyed and withered portion of those kidneys is represented in the foregoing illustrations (Figs. 13, 14, 15).

The destructive process is irregularly distributed throughout the kidney; but it may be broadly stated that it commences in the cortex and attacks the tubular portions later on; and that at an advanced stage the cortex is the seat of more pronounced change than the central portions.

Tube-casts.—It is this combination of disease in various stages in one kidney which, as Sir William Roberts points out, gives the varying casts found in the same case. 'The diversity in the character of the casts arises from the different condition of the several parts of the glands. In some portions the tubuli may be denuded of their epithelium, and the exudation thrown into them is discharged in the form of large hyaline casts; if the denuded portions have undergone subsequent contraction, the casts will be small and hyaline. Other tubes, clothed or partially clothed with epithelium, shed some of their cells with the contained exudation, and cause the appearance in the urine of casts more or less studded with epithelial remains. The longer the exudation is retained within the tubuli, the darker and more granular will it appear when discharged as casts; and *vice versâ*, casts speedily discharged are commonly hyaline. Sometimes casts are darkened by the colouring matter of the blood; and the opaque granular ones are (sometimes at least) composed of crushed epithelial débris moulded into the form of the tubule.'

The existence of such tube-casts in the urine tells unquestionably that interstitial nephritis is on foot; but they speak in broken accents only as to the extent of kidney destruction. Like every other semeion, they must be calculated along with the other phenomena of each case. When albumen is present in the urine, no one is justified in pronouncing the case to be one of 'Bright's Disease' until such tube-casts are detected; and if this were imperatively insisted upon by clinical teachers, much misery to patients and their families would be avoided; and also many a rent and tear in professional reputations. Still, even when unquestionable tube-casts are found, they only establish the fact that chronic renal disease is on foot; they cannot tell the extent of the disease, or give the measure of the destruction wrought.

Of course where some tubular nephritis co-exists with vaso-renal change, these tube-casts are more numerous than where uncomplicated renal cirrhosis alone is present. In the last the tube-casts are scanty, ' and it is not uncommon for them to be altogether absent for limited periods.' When carefully observed they furnish a clue to the nature of the morbid process going on in the kidneys. Granular casts tell of cirrhosis; while casts specked with fat reveal a fatty kidney. Hyaline casts, large or small, tell of chronic interstitial nephritis.

The accompanying engraving shows the hyaline and granular casts which are common in the vaso-renal changes. But even tube-casts have but a restricted

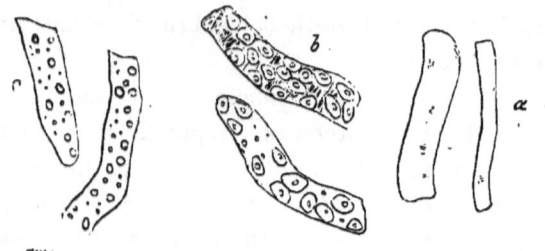

Fig. 16.—Tube-casts.—*a*, hyaline; *b*, epithelial; *c*, granular fatty. Magnification, 300 diameters.

diagnostic value; though their presence along with albuminuria certainly decides the question of kidney injury being present, either as cause or as coincidence. (In a case of intercurrent jaundice, the hyaline tube-casts were stained a deep yellow colour.)

The Urine.—That considerable modifications are under these circumstances produced upon the renal secretion, is only what could reasonably be expected. In the gouty type, or Norse victim of vaso-renal change, the bulk of urine is copious, and its reaction is acid. Usually, especially in the Norse type, it is clear. In the neurotic the bulk varies; is perhaps less copious than in the Norse type, and decidedly more given to throw down a sediment; while in others, again, the bulk is not markedly increased, and the specific gravity is from 1020 to 1024 or more. Certainly variations in character from the copious pale urine of low specific gravity to a concentrated urine of high specific gravity, are more marked in these persons than in persons free from this change. Of course such matters as pyrexia, diarrhœa, diaphoresis, affect the urine. A mitral lesion in the hypertrophied heart would profoundly influence the character of the urine. Traube's rule of a copious urine going with high arterial tension must form the basis of all calculations on the subject.

Dr. Garrod examined carefully the urine of a number of persons suffering from chronic gout; 'the majority at the time the examinations were made having no very urgent symptoms, but suffering from the sequelæ of this disease, as concretions of urate of soda in different parts of the body, and stiffened and deformed joints.' (This evidence of the uric acid formation will scarcely be challenged by anyone, however sceptical.) His conclusions are as follows:

'It is rather paler than healthy urine of lower density, and increased in quantity.

'The amount of urea, except in extreme cases, remains much as in health; the character of the diet being taken into consideration.

'The uric acid is very much diminished, and liable to be excreted in an intermittent manner. A small amount of albumen is very frequently present.

'The occurrence of deposits in the urine is not common; they occasionally occur during the cooling of the fluid, either in the form of urate of soda, or as rhombs of uric acid, more or less coloured.'

Such is the urine of persons of the uric acid formation of the Norse type in my own experience.

During attacks of acute gout—'the thunder-storms which clear the air'—he found the output of uric acid to be increased; but again falling as the attack passed off.

As to the urine passed in acute gout, Sir Charles Scudamore wrote: 'A deposition of pink or brickdust sediment on cooling of the urine is of such ordinary occurrence when any active symptoms of gout are present, that its connection becomes forcibly impressed on the mind of the patient, and he gives it the name of "gouty urine."'

The sediments found with the uric acid formation are rhombs of uric acid, looking like cayenne grains, which form after the urine has stood some hours; and red, pink, brickdust, or fawn lithates, the urates of soda and ammonia. The first is found with the Norse type, as a broad rule; the latter with the neurotic type. Oxalates are present at times, and often cause considerable irritation in the urinary passages.

Sometimes the output of urates is so copious, and the urine so laden with them, that the term 'a fit of the gravel' is applied thereto. They may be formations of various sizes, from what looks amorphous to the eye, up to 'sand,' visible particles; or 'calculi' as large as peas. Calculi form in the tubuli uriniferi; and kidneys sometimes show a large quantity of small uric acid concretions. One or more of these may fall into the pelvis of the kidney, lodge there, and wax larger. It may stay there as a renal calculus, producing its own phenomena; or it may become dislodged and pass into the ureter, causing more or less 'renal colic;' and then find its way into the bladder and be voided; or it may lodge in the bladder, and reside there as stone in the bladder,—requiring an operation for its destruction, or removal.

One of the curious matters of which the variations in the vaso-renal change furnish us with so many, is the localities haunted by stone in the bladder. It is notoriously present in the Eastern counties of Norfolk and Suffolk, and again in Aberdeenshire. While over other large areas vesical calculus is an unknown disease.

Uric acid has its companion oxalic acid; and calculi mainly composed of uric acid, or urates often exhibit layers of oxalates. Oxalates are common in the urine of gouty subjects. They are credited with possessing very irritant properties; whether as regards the brain and irritability of temper, or as regards the discomfort they produce in the urethra. The irritant urine of the uric acid formation may indeed set up cystitis or urethritis. The latter certainly is not uncommon.

When an advanced condition of renal cirrhosis is reached the urine is apt to be deficient in urea, to be pale and of low specific gravity, being 1010, or as low as 1005. Such cases, or rather such a condition, is of very unhopeful omen in my personal experience.

On the other hand, where there is a fairly normal urine, and times of increased output of urine-solids, 'the specific gravity may rise to 1030, or even 1040' (Roberts).

The bulk of the urine is a matter too little considered in connection with the progress of the vaso-renal change. While the blood-pressure within the arteries is well maintained, the bulk of urine keeps high. When the heart begins to fail the blood-pressure falls, and with it the bulk of urine. Sir Wm. Jenner has pointed out the grave significance of a fall in the bulk of urine. When in answer to the question, 'Do you pass as much water as you ever did?' the patient's reply is, 'No, I do not make as much water as I once did,' the answer tells that the patient has passed 'the turn of the hill,' and begun the descent. Of scarcely less sinister omen is the change betwixt almost always feeling too warm, and beginning to feel the cold. When the individual who is the victim of the uric acid formation complains of the cold, and contrasts this state with his previous indifference to cold, or rather even enjoyment of it, he tells of having passed 'the turn of the hill.' If the sensitiveness to cold co-exists with a gradual fall in the bulk of urine, then it becomes possible to forecast the future, and measure the span of life remaining; indeed, to realize that 'the spring is running down' fast.

Albuminuria.—When Dr. Bright made the observation that when along with anasarca the urine was albuminous, there was present disease in the kidneys—a perfectly correct conclusion—he probably little thought that half a century later the anasarca would drop out of sight in making a diagnosis. The detection of albuminous urine dates further back than the day of Dr. Bright; to whom, however, is due the credit of discovering the diagnostic value of the phenomenon. As to the precise import of albumen in the urine the greatest variations of opinion exist. It may truthfully be said that its significance became so grossly exaggerated, that a natural reaction set in. Medical practitioners have told me how strong was the hold of albuminuria on the medical mind some time ago, that when it was discovered, the patient was doomed straight off; and if medical men found it in their own cases, they took to their beds, and waited for the advent of the King of Terrors. But as the latter failed to call, after treading 'the valley of the shadow of death,' in time they took heart, left their beds and went back to their work. A few years later they grew to laugh at the bad scare they experienced. The fatal facility of the test-tube and its ready revelations have proved a temptation which many could not resist. The stress laid upon the reaction of the urine, the ominous significance of albuminuria at the examination table, have misled the profession and upset the mind of the public. An indiscreet medical man finds albumen in a patient's urine, and tells him he has got 'Bright's Disease.' The patient has an idea that 'Bright's Disease is incurable,' and that he is a doomed man. (Perhaps he may be; and Bright's Disease is incurable: and we must all die some day; and a goodly proportion of us will perish by some outcomes of vaso-renal change sure enough; but he need not be in such hurry about it!) The utmost alarm is excited, and a very distinct addition made to the sum-total of human misery. When the patient sees some other medical man who places the matter before him with greater regard to the clinical facts, the patient's mind is greatly relieved; but he feels very angry with the medical opinion originally given him. It is not to decry examination of the urine this essay is written; but one distinct factor in the group of motives for writing it is to enter a solemn and vigorous protest against the too prevalent practice of diagnosing 'Bright's Disease' merely because the urine is albuminous; without searching for other evidence of vaso-renal change. By presenting a fairly wide and comprehensive view of it, the indications furnished from the kidney will be, or at least can be, corrected by the other semeia of the widespread change. That the evidence so readily furnished by the test-tube may be weighed in the balance, and appraised accordingly. The medical observer once taught to see renal cirrhosis as but a fragment of a widespread change, will avoid rash conclusions and erroneous inferences based on an imperfect, and too often slovenly examination of the patient. This is rather strong language to employ; but its vigour falls short of the crying necessity for such outspokenness. It is not the knowledge of book-writers which is at fault—they enter their protests honestly enough; but the bent of the rank and file of medicine which calls forth these comments. It is enough to make one despair of the reasoning powers of one's species to contemplate an age which refuses to discriminate betwixt albuminuria and Bright's Disease; betwixt glycosuria and diabetes; and which persists in regarding beef-tea as a sustaining food!

Sir Thomas Watson faced the issue boldly, and with his wonted clear-sightedness. In discussing the subject of Bright's Disease, he wrote: 'Two questions of great interest at once present themselves:

'1. Does albuminous urine *always* imply the presence of Bright's Disease?

'2. Is Bright's Disease, when present, *always* accompanied by albuminous urine?

'To both these questions the answer is—No.'

Nothing could be fairer than this position. It is one simply unassailable. As to what extent albuminuria is found with the vaso-renal change, no one observer is in a position to say. Experiences vary. One man's experience does not tally exactly with another man's. One medical man will recount the cases where he formed a grave prognosis from the presence of albumen in the urine. Another will tell of the number of times he has seen albuminuria not followed by anything sinister. Sir Joseph Fayrer, a little time ago told me of a case where the patient's urine had contained notable quantities of albumen persistingly for twelve years. Yet the man was in good health.

We all recognise the fact that when a gouty man advancing in years passes a small quantity of albumen pretty steadily, we do not like the prospect. When the heart is failing and albumen begins to show itself in the urine we recognise it as the herald of disaster. But there are many circumstances where albumen is present in the urine which cause it to be a riddle not easy to read; and too often a puzzle of Sphinx-like character. As regards my own personal position in the matter, it is—that one casual examination of the urine may give its general characteristics, but cannot entitle it to be taken as a guide; and that repeated examinations extending over some time are requisite to careful diagnosis. As in a great many instances the patient can only give one, or at most two, opportunities for such observation, and as the urine varies more from day to day in persons undergoing the vaso-renal change than in perfectly healthy person, I have cast about for other semeia permanently present; and trust more to the grouping of symptoms, the *tout ensemble* of the case—not forgetting the urine—than to the examination of the urine, as the only safe infallible guide. My position is the outcome of my own personal experience; which must just be taken for what it is worth. One further word on this subject. It is much easier to detect albumen in the urine than to appraise its significance when found! There are circumstances where its presence excites no anxiety; there are conditions where its absence conveys no comfort. Frequently the evidence of vaso-renal change is so complete that it is a matter of comparative indifference what the reaction of the urine may be; if free from albumen it does not disqualify the other evidence; if albumen be present, that fact cannot add to the gravity of the case.

Certainly, when the urine is found to be albuminous no sensible or prudent man would dismiss the matter as trivial, or unimportant.

The necessity for repeated examinations extending over some considerable time is furnished by the fact that a New York Insurance Company has found the urine to present evidence of the presence of albumen in one in every eleven cases; while Leube found the urine of four out of 119 soldiers to present evidence of albumen in the morning, and no less than sixteen to do the same after a march. A solitary examination of the urine is a sorry guide to rely upon. And as two consultations are about the average of patients' visits to a London consultant; unless the said consultant has got some other evidence to guide him, and on which he can rely in forming his opinion, other than that furnished by the urine, it seems to me that the opinion would be of very little value to the patient.

The circumstances under which peptones find their way into the urine are scarcely yet quite clear to us; but we know they are apt to appear when the liver is out of order—and that is a common accident with persons undergoing the vaso-renal change.

Again, a dietary 'rich in albuminoids' is apt to cause albumen to be present in the urine; but 'rich' is a comparative term. What may be quite a normal amount of albuminoids for persons of the Norse variety may readily enough be 'rich' to another of the Neurotic type—'rich' enough to cause the urine to be albuminous!

The subject of albuminuria and the significance of albumen in the urine is not a matter to be dismissed summarily. And the writer trusts that what has been written here *pro* and *con.*, will be 'read, marked, learned and inwardly digested' by the reader as worthy of consideration; and having a direct bearing on practice. And let him not forget that in Quain's 'Dictionary of Medicine,' Dr. Lauder Brunton, F.R.S., draws the distinction betwixt 'true' and 'false' albuminuria.

Glycosuria.—It is now a well-recognised clinical fact that persons of the uric acid formation are liable to pass quantities of sugar in their urine. That the liver which reveals its incompetence by reversion to the formation of primitive urine-stuff, should manifest derangements as to its glycogenic arrangements, need occasion no surprise to anyone, to my way of thinking. Garrod has observed that gouty persons are liable to pass saccharine urine; and the same observation has been made by others. It certainly falls in with my own experience. A considerable number of gouty persons pass sugar, to my personal knowledge. Just as emotion affects the lachrymal glands, arousing their activity, and the salivary glands by arresting their action, so does it affect the liver; as seen in the production of jaundice by mental shock, or surprise, and in acute diabetes from like causes. The production of glycosuria, often deepening into diabetes by carking care, or prolonged anxiety, from business or other cause, is now widely recognised. C. Creighton holds that persons who do not betray emotion readily, and who carry themselves stoically under trials, are especially liable to have their livers deranged; and some other trustworthy observers hold the same view. True diabetes seems to me to have a special affinity for persons of the neurotic type, while glycosuria is common with persons of the Norse type. When the general nutrition does not suffer, the case may be classed fairly enough as 'glycosuria;' where the general nutrition suffers, then the term 'diabetes' is correct. A glycosuric person may be liable to acute 'diabetic storms' from time to time; as is not unfrequently the case.

Glycosuria is linked with the uric acid formation by other ties than the fact that persons of this formation are liable to glycosuria. There is another matter to be considered. The person with glycosuria is often put upon a dietary rich in albuminoids; this was a practice of the French school a generation ago, and still too commonly prevalent. In keeping the vision directed to the prevention of sugar-formation too exclusively, the fact of the excess of albuminoids, and the consequences thereof, is apt to be lost sight of. Yet this is not *une quantité négligeable* by any means.

While it is perfectly true that many robust gouty persons pass quantities of sugar; and that, too, not always as a transient affair, but persistently for long periods without any ill effects resulting therefrom: it is equally a clinical fact that old subjects of the vaso-renal change often burn out at last with saccharine urine. They have worn well and are well stricken in years, when they begin to lose flesh, and sugar is found in the urine; and in a little time it is all over with them.

In glycosuria, as in albuminuria, rash conclusions are to be deprecated as apt to be misleading. Either condition ought to evoke careful, thoughtful consideration. In the matter of glycosuria, its mode of origin, its march, and the circumstances of the individual should all be rigorously investigated; and the information elicited be made the subject of thoughtful ratiocination. If the patient be a neurotic individual, giving evidence of the vaso-renal change, and looking older than his years, in my opinion, the matter of his burden of care or worry—how far he can escape from it; how far it is fastened upon him inextricably—is not a factor to be lost sight of, or left out of the calculation as to the prognosis of the case.

In fact it is often possible for the reflecting physician to be of much practical service to his patients, beyond merely treating their passing maladies. If he possess a competent acquaintance with the long chronic morbid process (some account of which has been given in these pages), he will often be able to detect the oncoming of disease—long before its progress is obvious to the untrained eye. An active business man comes to him complaining of 'not being up to the mark;' being listless in mind and body, and losing flesh. An examination of the urine reveals the presence of sugar. The note of alarm is at once sounded. Not, as in a case reported to me some time ago, where on such discovery the medical man hastened to the patient's wife, and abruptly told her that her husband had got 'Diabetes.' With as little consideration, in turn, she told her husband what the doctor had said; upon which the poor man exclaimed: 'Then I am a dead man!' and in less than a week he drew his last breath. Instead of such brutal thoughtless behaviour the reader of this essay will, it is my earnest hope, break the matter gently to the patient—as is his duty as a neighbour and a man; explain to him the facts of the case, and the necessity for some rest for cure; and some lightening of the load he carries for the future. To make him understand that it is no more inevitable that a case of glycosuria will march steadily on to the grave, by the route of confirmed diabetes, than that a person with incipient phthisis shall succumb to consumption; or a man in temporary financial embarrassment will of necessity become a bankrupt. Nor when a neurotic is complaining of indigestion, acidity and flatulence, losing flesh, passing lithates, and subject to sharp neuralgic-colicky pains across the transverse colon, to diagnose inflammation, and inject morphia; by which the incapable liver is further depressed; and then try to feed up the sufferer by plenty of animal food which is just so much poison to him—as happened to a New England patient of mine lately. When he finds such a person becoming sleepless he can warn him of the coming trouble ahead of him—if he does not slacken the pace at which he is going.

We are beginning to realize what the old physicians call the *pathemata mentis*, the effect of mental conditions upon visceral conditions; of the reaction of the tissues of the epiblast upon those of the hypoblast; and *vice versâ*. Not only does the overworked man injure his own assimilative processes, but he impairs those of his children. 'What is acquired by the father is inherited by the child;' and this does not apply solely and merely to worldly goods. Again it is a case of the fathers having eaten sour grapes and the children's teeth being set on edge. One American lady after another, complaining of digestive troubles, and exhibiting evidences of the neurotic aspect of the vaso-renal change, has told me of the energy displayed by her father, for a long period of years; until the impression has been forced upon me that the two have something more than an accidental association. And this small fact stands in a certain suggestive relation to the great fact that the inhabitants of the United States of America are at once active-minded, industrious, and dyspeptic. It is not merely their dietetic outrages as to choice of food, and the little consideration they give the stomach by taking their meals in haste, instead of in a leisurely rational manner; which give primary gastric derangement. And the person reaps as he sows! The long mental strain is not borne with impunity. Nature takes her revenge upon the liver. The hard-working man not only impairs the efficiency of his own liver; but his children come into the world with insufficient, or incapable livers. The liver governs the appetite for food; and consequently such beings are naturally and instinctively small eaters. But mankind is impressed with the idea that every day so much food ought to be taken—as one of the fundamental laws of the universe—and that if a child be delicate it ought to be 'fed up.'

Old Dame Nature, in her kindly thoughtful way, has made the child a dainty fastidious eater, with a small appetite—as a protection for the feeble liver; but in vain! Her purpose is ruthlessly thwarted; with the best intentions, doubtless! By precept the child is encouraged to eat more than it wants; while it is tempted by delicacies of all kinds. Nature is trying to level down the physique to the liver; and when her plans are rudely traversed, she resents the outrage; and pays it out by liver-reversion, and the formation of uric acid. Now the intelligent physician, perceiving the real facts of the case, will be able to interpose on the poor child's behalf; just as a man would in any other relation in life when a child is being oppressed—and stay their tyrannical efforts. Of course this will be more difficult at first than when the lay world has become familiar with this matter; and the fact that vaso-renal change is to a great extent an outcome of modern life, is generally realized.

It is needless to multiply examples of how the physician may be able to help his fellow-creatures by certain knowledge; and feel his hands strengthened, not only to cure but to save and prevent, by familiarity with morbid processes extending over long years. Even something more is possible; he may be able to correctly estimate a child's inherited physique, and so wisely guide its upbringing, and control its rearing, as to enable it to grow in a favourable manner; and so to lessen, and not intensify its inherited proclivities.

Indeed, he may so digest and arrange his knowledge as to make it available in the way of guiding men and women, as to their own ways and modes of living; and, more than that, of correcting the inherited tendencies of their offspring. And this is only the more necessary as the tendency to move in the direction of the neurotic is waxing stronger day by day; especially amongst urban populations. As to the subject last discussed, viz., glycosuria, our acquaintance with the malady is now such that infinitely more valuable advice and suggestions can now be given to a patient, than was humanly possible twenty years ago; and this chiefly by recognition of the reaction between epiblast and hypoblast—between the brain, as the organ of mind, and the glandular apparatus of the assimilative organs.

Results of Toxic Blood.—There are a series of morbid manifestations in persons who are the subject of vaso-renal change, which take their origin in waste-laden, or imperfectly depurated blood. Whether this last is due to excessive formation of abnormal rather than normal products by the liver to some extent, with imperfect excretory action in the kidney; or the latter alone exists, is a matter of secondary moment. The blood is laden with products of albumen-metamorphosis—that is the cardinal fact. It is when the kidneys become distinctly inadequate, and waste matters accumulate in the blood, that we see the most pronounced evidence of vaso-renal change being a long, persistent, sustained, and often desperate attempt of the system to depurate the blood. These morbid manifestations affect the brain, the serous membranes, the mucous membranes, the lungs and the skin.

Uræmia.—This is the term applied to the condition when the sensorium is affected. Headache is common—sometimes it is hemicrania; or in another individual the pain is felt across the forehead, and at times at the occiput or nape of the neck. Very frequently the pain is vertical. The vision is influenced; being more or less completely lost at times. The hearing is less commonly affected. Drowsiness is often present. There are twitchings of the voluntary muscles. When a more intense condition is present, there are convulsions of the epileptic type with coma. Very often such a condition is accompanied by vomiting; the vomited matter either being ammoniacal, or of a urinous odour; and soon undergoing ammoniacal decomposition. If a microscopic slide be moistened with hydrochloric acid and held under the nose, crystals of hydrochlorate of ammonia are quickly formed. The condition of uræmia is a very serious one, fraught with great danger to life when convulsions and coma appear; while death from effusion into the brain is a common termination. At other times the paroxysm is survived. The condition varies; sometimes the insensibility betwixt the convulsions is complete; while in other cases the drowsy sufferer can be roused. In a case seen in the Hospital of Vienna, the unconsciousness deepened, and then lifted at short intervals; so as to remind one of a mist rising, and settling again. The condition of uræmia may alternate with that of some other outcome of the toxic state of the blood.

Secondary Inflammations.—These may attack the serous membranes, and frequently threaten life; the pleura, the peritoneum, the pericardium, or the meninges of the brain being affected.

Then the mucous membranes feel the effects of the waste-laden blood. Bronchitis is common; the sputum being distinctly acid when the affection is chronic. At other times there is muco-enteritis; with the discharge of fluid of urinous odour, and readily undergoing ammoniacal decomposition. This diarrhœa is of a compensatory character, and should not rashly be interfered with. The late Dr. Carpenter, in his article on 'Secretion' in the 'Cyclopædia of Anatomy and Physiology,' points to the fact that each excretory organ is capable of supplementing the function of another organ, when defective or embarrassed; and eliminating material special to the other organs rather than itself: a property which they possess as mere modification of the general excretory surface of the lower forms of life. Thus the kidneys cast out bile in jaundice; the bowel urinary salts in Bright's disease. When such diarrhœa is present, it is well to bear the fact in mind; and *not* arrest the diarrhœa until the action of the kidneys is established. By so doing in my early days of general practice, an old lady, the subject of vaso-renal change, was tided over an attack of uræmic diarrhœa which gravely threatened life; though my masterly inactivity only procured me some odium— my motives not being comprehended by my critics.

Inflammation of the lungs is certainly often secondary, and due to the toxic properties of the blood. In one such case seen in early days, where the urine was distinctly bloody, the pneumonia readily yielded to a mixture of potash with nitric ether and juniper. Nor is such a case exceptional by any means.

The skin is liable to prurigo (as in jaundice), to impetigo, herpes zoster, and eczema. Erysipelas is not rare: it may extend over a lower extremity, or be confined to the face. The skin becomes thin—in some cases almost like a sheet of paper, or parchment; and this is very conspicuous when distended by anasarca. In one such case seen a few years ago, after an attack of lung-inflammation (set up by a long-continued fog) the legs swelled. The gentleman, an old victim to gout, was severely tried by the lung-trouble; and when an erysipelatous blush spread over one leg, with one dark spot about the size of a florin, which seemed as if about to slough out; Sir Joseph Fayrer and myself quite thought the end was at hand. But, despite the unfavourable circumstances, the old gentleman rallied and recovered.

The great matter to remember in the treatment of conditions secondary to vaso-renal change, is the condition of the blood; which is the exciting cause. Instead of following the treatment appropriate to primary disease of like character, the most effective measure is to sweep the waste matters out of the blood as speedily as may be. This can be done by alkaline cathartics and diuretics pushed freely; while the skin is thrown into action by external heat and moisture. But the practitioner must know what he is doing in order to act aright. The great necessity for comprehending how 'secondary' inflammations differ from 'primary' inflammation was unfortunately seen lately. A gouty gentleman had acute articular gout, which was suppressed; with the result of peritonitis. The metastasis never struck the medical man in attendance; who pushed opium, with a fatal result.

As to the complications found with the vaso-renal change, they are well illustrated by the accompanying table of four hundred cases collected by Sir William Roberts, M.D.:

Heart.	Lungs.	Pleura.	Pericardium.	Peritoneum.
Hypertrophy, 125 H. with valvular disease, 54	Œdema, 115 Pneumonia, 52 Pn. Apoplexy, 8 Gangrene, 4 Tubercle, 37 Vesicular Emphysema, 33	Pleurisy, 57	Pericarditis, 30	Peritonitis, 45

Liver.	Spleen.	Stomach and Intestines.	Brain.
Cirrhosis, 41 Fatty liver, 38 Nutmeg liver, 11 Lardaceous, 3	Chronic tumour, 58 Acute splenitis, 17 Contracted spleen, 9	Gastric catarrh, 36 Catarrh, with follicular ulceration of intestines, 85 Tuberculosis of intestines, 13	Sanguineous apoplexy, 14 Effusion of serum under arachnoid, 59 Tumour cerebri, 11 Chronic arachnitis, 6

This gives at a glance the morbid outcomes of the vaso-renal change. It is somewhat remarkable that there is no account of aneurysm.

Gouty Asthma.—The effect of uric acid upon the respiratory centre is interesting. Girls with the uric acid diathesis are liable to attacks of bronchial spasm whenever there is any bronchial irritation; and such a condition is common in the first stage of acute bronchitis. At times there is dyspnœa—certainly not the result of any mischief in the lung, and clearly neurosal. The neurotic is liable to distinct attacks of asthma. Then, of course, the old gouty man with chronic bronchitis and emphysema, has times of distinct bronchial spasm. (He has, however, also acute attacks of difficulty of breathing, from the sudden evolution of gas in the intestines preventing the descent of the diaphragm.) Sometimes there is a distinct embarrassment of the breathing, lasting for several days. One such experience befell myself two years ago. After lasting two days, a sharp action of the bowels occurred about five o'clock in the morning—the contents of the evacuation being highly bilious. A sound sleep followed; and, on awakening at the usual hour, the embarrassment of the breathing had so entirely disappeared that it was difficult to realize that it had been so recent.

Nocturnal dyspnœa is not uncommon. It seems as if the respiratory centre partially slept; and an accumulation of carbonic acid in the blood roused the centre to great activity. After a little energetic or hard breathing, the condition of affairs is rectified; and then sleep soon follows. This has been noticed by Basham, who terms them 'inexplicable fits of difficult breathing.' They differ from attacks of true cardiac dyspnœa in being altogether less severe and sooner over.

Dupuytren's Contraction.—This is a growth of connective tissue in the palm, over a flexor tendon—usually that of the ring-finger. It extends to the sheath of the tendon, and gradually brings the finger down. Sir James Paget has traced its connection with gout; and certainly it has a high diagnostic value in times of doubt. It usually occurs in the right hand; and is attributed to the walking-stick, or umbrella-handle. Possibly enough, these are the exciting causes; but there is something else as well. One old gentleman who had it in his left hand explained its presence as being due to his geological specimens; which rested on his left palm while he exercised his hammer upon them. That there are exciting causes in action may at once be admitted: but then there is something lying beneath which permits these causes to operate; and that evidently is the uric acid diathesis. Outward applications of iodine do no good; and operative proceedings are not satisfactory, as the results are temporary.

Albuminuric Retinitis.—Another curious local growth of connective tissue, in the vaso-renal change, takes place in the retina. R. Marcus Gunn has provided me with this brief account of it:

'It is most commonly associated with chronic renal disease; especially the contracting form. It has been observed in about thirteen per cent. of all cases of albuminuria. Ophthalmoscopically we find numerous soft-edged patches, mainly in the neighbourhood of the optic disc, and behind the renal vessels; and minute opaque, very white dots or striæ at the yellow spot, arranged in the form of an asterisk, with its centre at the fovea centralis. The large spots nearer the disc occur earlier than the dotted arrangement at the maculæ. Besides these more typical retinal changes, we may also have numerous hæmorrhages in the nerve-fibre layers; especially in cases where the papilla is much swollen. Not infrequently, too, the retinal arteries are unusually conspicuous, from their walls reflecting more light than do those of healthy vessels. Both eyes are always practically affected.

'Pathology.—(1) *Changes in vessels:* There is sclerosis of the walls of the smaller arteries. The larger vessels exhibit less sclerosis; but their adventitia is often markedly hypertrophied. (2) *Changes in the retina:* The large soft-edged patches are principally due to the deposit of cells loaded with fatty granules in the middle layer of the retina. The minute dots at the maculæ are caused by fatty infiltration of the inner ends of the Müllerian fibres in this situation. Besides this fatty change in the radial fibres, they are found to be both elongated and thickened; and sometimes to have papillary outgrowths of connective tissue springing from them. Frequently the nerve-fibres on the disc, and in its immediate neighbourhood, undergo a varicose sclerosing hypertrophy.'

We see, then, a growth of connective tissue, both in the retinal layers and the walls of the small arteries. And both here and in Dupuytren's contraction, we see marked illustrations of that growth of lowly connective tissue; which was laid down in Chapter II. as the anatomical departure *par excellence* of the vaso-renal change.

The middle stage extends over a considerable period; being established in some instances at about the age of forty-five; and passing into the final stage shortly after sixty. In my own experience this rapid career has taken place in licensed victuallers mainly. In other cases the middle stage is not well established till the age of sixty; and the final stage sets in about seventy-five. But in some families the breakdown is later still; and the spring runs down very slowly.

CHAPTER V.

ADVANCED STAGE.

General Considerations—Lead—Occlusion of the Coronary Vessels with Fatty Degeneration of the Heart—Arcus Senilis—The Descent—Venous Fulness—Interstitial Changes—Dropsy—Albuminuria—Serous Effusions—Death, Sudden or Slow.

PROLOGUE.

AT this point some general considerations seem called for, which can now be discussed with advantage.

Without a certain familiarity with the widespread change sketched in the preceding chapters, it is difficult for any medical man to grasp the entirety of a case; when some particular outcome of this change is the immediate cause why the person of advanced years calls in medical aid. It may be an attack of bronchitis, or may be pleurisy, or diarrhœa, or eczema, or angina pectoris, or neuralgia, or gout; it matters little what, it is only the prominent feature of the case—that portion of a great widespread and invisible morbid change which is visible. A great many matters are apt to remain shrouded in darkness, if the associations of the particular malady are not clearly recognised. The medical man is called in to a case of apoplexy. If he find a large heart, and a hard artery; and bear in mind what Rokitanski observed; he will see that unless he can promptly lower the blood-pressure in the arterial system, the clot will wax larger and larger; with the result of ultimately squeezing the life out of the medulla—and with that the existence of the individual will terminate. The only method of doing this is to bleed the patient within an ace of his life; for he cannot possibly be in any greater danger from excessive venesection than he is from the growing clot. In one such case which came under my notice in South Wales, moderate bleeding was tried; but without avail. It must be carried out intelligently and heroically. Nothing else offers even a prospect of success. The pressure within the cerebral artery must be reduced to a level with the resistance offered by the contents of the cranium; otherwise the clot must necessarily wax larger. The patient can but die.

The patient developing chronic bronchitis will tell of the bulk of urine falling off since the bronchitis came on. What information does that item afford us? It tells, in unmistakable accents, that the right ventricle is embarrassed. Percussion can elicit nothing. Emphysematous lung more or less completely covers the heart; perhaps it is impossible to make out any cardiac dulness worth the name. Physical examination is dumb. There are ossified costal cartilages, rigid and resisting; and, behind them, distended emphysematous lung-borders; how can we get at the condition of the right ventricle? Possibly by auscultation; a disproportion may be discovered betwixt the first sound of the heart, and the pulse at the wrist. The large right ventricle furnishes a good first sound, audible even through the rarified lung-tissue—a bad conductor of sound. While the left ventricle sends on such an amount of blood as comes over to it from the right heart, *i.e.*, but a small quantity, not producing much distension of the arterial system. Still that is not very valid evidence. But if the fall in the bulk of urine be such as to attract the patient's attention, and he speaks confidently, it tells, in no broken accents, as to the condition of the right ventricle.

And here a short digression seems desirable for the sake of junior readers. At the hospital great reliance is placed on physical examination : and little on statements made by the patients, or their friends. The class of persons who are encountered at a hospital consists, in the main, of people who have had something else to do in this world than minutely note their subjective sensations; their attention has been called elsewhere by the exigencies of their situation. If they have noted them they lack language in which they can express themselves precisely. Probably, too, they have not studied accuracy, or precision in description, as a habit of life; and so mislead or give incorrect answers, without designing to do anything of the kind. No one who has had much experience in a hospital is likely to challenge this statement. When the ex-student or hospital resident encounters private patients, he finds to a large extent a very different class. First of all they are better educated, and can express themselves correctly; and often fluently. Many have not had to work for a living, and, having no great calls upon their attention, have studied themselves, and their feelings, minutely. A number, too, have been more or less invalids; and have preserved a careful record of their experiences. Fishing the seas of the patient's sensations is very unprofitable with hospital-patients; but in the case of an intelligent private patient it is far different; and the most valuable information may often be volunteered, or elicited by dint of careful questioning; and sometimes cross-examination. A mine of matter, indeed, can often be found with patient persistence. And this applies especially to the intelligent lady who is often the victim of some outcome of vaso-renal change. But there is much in the manner of putting questions to such a patient. She has a highly developed nervous system, and is remarkably acute; so that she will perceive the drift of any interrogation at once, almost before the last word is spoken. If she observes that the questions hang together, she will give willing aid to the questioner. If her conclusions be unfavourable, the well-spring of information will dry up promptly. If the questioner can so put his questions as to reveal his thorough acquaintance with the subject, he gains her confidence forthwith. But if they but betray unfamiliarity with her malady, or complaint; and give her the impression that they are no more than curiosity, she will as certainly withhold her confidence. For instance, we have seen that vesical irritability is a common accompaniment of migraine; whether in its complete, or its larval form. Now, if a comparatively young man has to make such an inquiry of a sensitive lady, everything almost will turn on the manner in which the question is asked. If it is made in such a manner as to convey the impression that it is a matter on which an affirmative answer is expected, and looked for; it at once strikes her forcibly as something linked with the other phenomena. If asked in an unfortunate manner, it will be regarded as simply impertinent; and as such be resented—and promptly.

Nothing impresses the patient so profoundly as a series

of well-planted questions, like arrows in succession striking the target. It is quite natural that this should be so. The questions tell of familiarity with the general condition; while questions which fly wide of the mark soon arouse the patient's suspicions. An acquaintance with the general condition of elderly patients will and can alone enable the medical man to show his grasp of the case in its entirety.

An illustration of what is meant is furnished by the following experience. Talking on matters medical to the lady friend of an old patient, the writer used the expression: 'The Almighty writes a legible hand when once we have learned to read it.'

The lady said, 'What do you mean?'

Gazing steadily at the bright neurotic, he asked: 'Madam, have you pain at one side of your head, and in the eye at times, with sparks of light?'

The reply was in the affirmative.

'And have you indigestion?'

'Yes!'

'With acidity and flatulence?'

Again, 'Yes.'

He then proceeded: 'I will tell you something. You have great irritability of the bladder with the headache.'

The lady looked much surprised, but replied firmly: 'Yes, I have.'

'Now, madam,' he continued, 'I will tell you something more. You have sediments in your water.'

Perceiving that the speaker was perfectly familiar with his topic, she answered: 'Yes, I have that, too.'

He then went on: 'Now, madam, do you understand what is meant by the expression, "The Almighty writes a legible hand when we learn to read it"?'

The lady understood perfectly after that. She then gave an account of other members of the family, who also suffered from migraine.

Now the manner in which these remarks were said, made all the difference betwixt a very aggravated form of impudence, and an interesting conversation devoid of offence; and making a strong impression upon the lady's mind.

Then another matter relates to prognosis. A great difference exists betwixt private and hospital patients as to their hold of life. The hospital patient has (probably) suffered from privation, led a hard life, very often an irregular life; there is often a history of both alcohol and syphilis—two highly important matters in prognosis—so that his, or her hold on life is greatly weakened; and the system easily succumbs. On the other hand, the private patient is often (of course not invariably) a person belonging to a long-lived family, a person of regular habits and abstemious ways; quite a contrast to the other. The one, in short, has a 'broken constitution;' the other has a 'sound constitution.' In each may be found a large heart, a hard artery, with albuminuria and tube-casts. But the two are not on a par; far from it! The prognosis in the one is readily made out, but in the other, it is very easy to trip. Further, errors in prognosis are apt to be remembered to the doctor's detriment; and the man who survives a gloomy prognosis—and in doing so disproves it—is apt to be given to talk about the subject. Indeed, many persons exult in recounting 'how they deceived the doctor.' Sir Andrew Clark, M.D., has uttered many a vigorous sentence of epigrammatic character, but never a better one than this: 'Do not apply to your private patient at the beginning of his disease the prognosis which is correct enough in the case of a hospital patient at the end of his.' And above all things, do not be led into the temptation of fixing the span of life. This is like a besetting sin with some men; and is usually a most dangerous practice for their reputations. Of all forms of mistake, the most gratuitous of all is—prophecy! Of course there are times and circumstances when the attempt must be made; but it should always be done with considerable reserve. 'Fools rush in where angels fear to tread,' certainly applies to a time-prognosis in a patient's bedroom very often.

After this digression the strictly medical aspect of the subject may be resumed.

Lead.—The relations of saturnine intoxication to gout has been a matter of observation for more than a century. Persons of the uric acid diathesis are peculiarly sensitive to lead; while persons who absorb lead are specially liable to the uric acid formation. A foreman workman, whose special business it was to mix paints, had very curiously deformed hands and feet from the deposits of lithates; furnishing a striking example of the relations of lead to gout.

Sauvages described 'lead arthralgia.' Dr. Parry, in 1807, spoke of 'Gout from Lead.' Dr. Garrod observed the effect of lead in diminishing the output of uric acid. His conclusion is as follows: 'It would appear, therefore, that in individuals impregnated with lead, the blood becomes loaded with uric acid; not from its increased formation, but from its imperfect excretion; and this is of much interest in connection with the fact that the subjects of lead-poisoning are *cæteris paribus*, more liable to be affected with gout; and as we shall see further on, that those who inherit the gouty diathesis are more likely to become poisoned by the imbibition of lead.' The administration of lead to animals for some length of time sets up interstitial nephritis of a progressive character.

It will often then be a desirable matter to look at the patient's gums. If the blue line be present it speaks for itself. It is less conspicuous when the gum is retracted; but this last fact has got a significance of its own.

Advanced Stage.—As pointed out at the end of the last chapter, the middle stage is completed sooner with some persons, and in some families, than in others. Sometimes this stage is prematurely brought to a close by some illness. Exposure to weather may start up bronchitis, or severe sciatica, giving the constitution a shake. One case is well known to me in a hale medical man of fifty-eight, belonging to a hardy race which made old bones, who caught typhoid fever, but was not laid up, or would not lay off work for it. This seemed to give a terrible impetus to the senile changes on foot; and instead of living fifteen years longer, as looked very likely, the spring ran down in thirteen months. A secondary serous inflammation may mark an epoch in the life of an individual. A not unusual expression used about a man after a serious illness is this, ' How it has aged him,' and especially when the person is past middle age. It may be mental worry or much affliction; but the result is the same—the spring runs down apace. In some others who are wearing well, and looking younger than their real years, the illness will bring their appearance on and up to their actual years; after which they proceed at the usual and normal rate.

'A man is as old as his arteries,' is a French expression often quoted by Dr. S. Wilks. And it is certainly so. No matter how youthful the other tissues, if the arteries are in an advanced stage, life is thereatened by arterial rupture on effort, or rise in arterial tension. The cases of sudden death, now so constantly attributed to 'heart disease,' are, in a large number of cases, rupture of a diseased artery, with or without a pre-existing aneurism. (The present fashion of attributing sudden death to heart disease reminds one of the habit of attributing sudden deaths to apoplexy before the diseases of the heart were known.) The effect of syphilis upon arterial change must be borne in mind, with its tendency to thicken the arterial inner coat (*tunica intima*). Such rupture may not always be due to effort: but at times to some arterial spasm raising the blood-pressure in the arteries. Rupture of an artery is certainly linked with sudden spells of cold weather.

Changes in the Vascular System.—As years wear on, the danger of life from alterations in the vascular system increases steadily. The large heart may burst a brittle artery, and often does. These are the rocks of Scylla. But beyond this risk, lies the deadlier terminal danger —the shoals of Charybdis, viz., 'fatty degeneration of the heart-wall;' the last act in this long pathological drama. Its march may be slow; it may give little or no sign; like the avenging deities, its feet may be shod with wool; its oncome may be insidious; but it will be present in time, sooner or later,—if the individual do not die in the meantime of something else. There is no escape from its clutches.

While making this statement I wish to express as

distinctly, as forcibly, and as unmistakably as lies in my power, that while fatty degeneration of the heart, as ordinarily understood, is a mortal disease, it is essentially a senile change; and indeed the last act of this long pathological process. As a matter of strict pathological fact it is found after severe pyrexiæ in an acute form, or pernicious anæmia or phosphorus poisoning; but, as a great clinical fact, it is associated with the gradual failure of the once large heart of the vaso-renal change; and, as such, has its morbid kith and kin, and its own pathological associations. And without these, the diagnosis of fatty degeneration is as cruel as it is unwarrantable by the clinical facts. A heart may have but a feeble impulse, and a weak first sound, and be in an ill-nourished and flabby condition, like any other ill-fed muscle; but to apply the term 'fatty degeneration' to such a state, is often simply diagnostic slovenliness. When the atheromatous change in the arteries is advanced, the change involves the vessels of the coronary circulation; and then the fate of the big heart is sealed. The atheromatous change is often more pronounced in the coronary arteries than elsewhere. They are forcibly distended by exposure to the aortic recoil (the old view), or to the cardiac systole (the view now accepted). But, all the same, I would remind the reader the diagnosis can only be made in the post-mortem room; there is no infallible evidence in life; at best it is but a suspicion,—though at times no doubt the suspicion is very strong indeed, and all but unavoidable. Nevertheless the term ought to be restricted to the deadhouse whence it originally emerged. The term, indeed, is one of those left-handed services which pathology has rendered to clinical medicine; it had a certain fascination about it; it was used when the causation of fatty degeneration was unrevealed to us; it was recklessly misapplied, and has been the source of incalculable human misery of avoidable nature. It has been indeed the cause of measureless unhappiness from its misuse.

In order to have clear views about fatty degeneration, a glance at the subject historically will be desirable. Laennec distinguished between fatty infiltration and fatty degeneration of the muscular fibre of the heart. The facts remained with very little advance made until Rokitanski took the subject up. In 1848, Sir Thomas Watson says of fatty degeneration, 'Walls thus soft are liable to yield under pressure; but I know of no particular symptom by which we can detect such a state of softening.' Yet forthwith the profession, and especially the less competent members of it, began to make the diagnosis of a disease of which no symptom in life was recognisable; according to one of the best observers who ever stood in the front rank of the profession. Great quietude of mind and body was thought essential to existence; and within my recollection men, advanced in years, walked about with the utmost circumspection, afraid of shock or fall, where the diagnosis of 'fatty degeneration' had been made by some one who knew little more of the disease than its name. (Dr. Fred. Roberts informs me that this kind of diagnosis is not yet extinct.) The impression upon the minds of the profession seem to have been something like this: 'The muscular structure of the heart is liable to degenerate into fatty débris (that had been fully demonstrated by several observers, especially Dr. Omerod), but under what circumstances was unknown; but the belief seems to have been that it was an idiopathic decay of the heart-wall of mysterious and unknown origin. Any view more calculated to sow unhappiness and misery in the human breast it is impossible to conceive. And right and left men set to work to diagnose fatty degeneration of the heart in the most reckless manner.* A feeble impulse and a faint first sound; and the diagnosis was made. One lady has told me of such a diagnosis being made in her husband's case thirty years ago, and of her being told 'never to leave him, as he might die at any moment' (though how her being beside him was to affect the result is not very apparent); the said husband being at the present time a hale man without the slightest evidence of any weakness about his heart. This is only one solitary instance out of many; and for a quarter of a century medical men went on in this fashion causing untold, incalculable misery; and in the end discrediting their own reputations. When Dr. Walshe brought out his work in 1851, he gave an indistinct apex beat, 'a feeble toneless first sound,' infrequent or irregular pulse, incapacity for exertion, fits of dyspnœa, angina more or less perfect, a tendency to syncope, and vertigo, as its main indications. He also wrote, 'Fatty disorganization of the heart is by no means necessarily fatal. I have known extensive destruction of the kind exist, where death has occurred from unconnected chronic diseases of other organs.' All which tell that by that time some real knowledge was being acquired on the subject.

Rokitanski first put the actual pathology of fatty degeneration on a sound basis. He noted its association

° Just as they did 'an atheromatous state of the cerebral arteries when the phrase came into use.

with 'ossification of the coronary arteries,' and also that it 'occurs more especially in hypertrophied and dilated hearts.' He describes it as presenting many centres scattered throughout the heart substance; where 'the muscular substance is pale, flaccid, of a dirty-yellow colour, and soft and friable;' while 'a microscopic examination shows an accumulation of black and dark outlined globules, which prove to be fat; while the muscular fibres are found to have lost their striated appearance, and the fibrillæ are soft, and readily break down into delicate molecules.' Again, he refers to its association with the enlarged heart : ' This form of adiposity most commonly occurs in the muscular substance of the left ventricle, and in cases of hypertrophy.' So that already fatty degeneration is looming up as the end of the large heart. He also remarks that rupture of the softened heart-wall by laceration, though it might be supposed to be frequent, yet ' such is very rarely the case.'

When Dr. Stokes of Dublin brought out his classical work on the 'Diseases of the Heart,' he discussed the subject at considerable length; accepting Rokitanski's description of its pathology. The association of the necrobiotic change with an atheromatous aorta begins to gleam through the recorded cases he gives; and therefore we are not surprised at the following being among his conclusions. It is associated with ' hepatic or renal disease; and, lastly, atheromatous alterations of the aorta. That, though often associated with the gouty state, it may occur independently of that condition. That the disease may, on the one hand, affect a heart already in a state of hypertrophy; or, on the other, of atrophy. That in the earlier stages but little change is apparent to the unassisted eye in the anatomical condition of the ventricle; and it is by microscopical examination alone that we can determine the actual freedom of the organ from disease.'

Here is one of the ablest physicians who ever breathed, stating in unmistakable language that microscopical examination of the structure of the heart alone could determine the presence or absence of fatty degeneration of the muscular structure of the heart; while on all sides men of ordinary parts, and by no means remarkable either for their clinical acumen or their familiarity with pathology, were making the diagnosis in an off-hand fashion; scattering misery and unhappiness broadcast over the land. It is no uncommon experience to find a patient telling that so many years ago his heart was pronounced to be in a state of fatty degeneration; a diagnosis utterly unjustifiable—and disproved by the subsequent history of the individual.

When about it, it may be well to thrash this matter out thoroughly; and knock the brains out of the demoralizing superstition that fatty degeneration of the heart is an idiopathic disease of muscle of mysterious origin; springing up like a thief in the night; slaying its victim without giving a hint of its presence: the dread and terror—the ghastly spectre of hundreds of unfortunate persons, who have been told by their medical advisers that they have fatty degeneration of the heart; or have made the diagnosis for themselves by dipping into medical literature of a date anterior to fairly full acquaintance with the disease. What says Dr. Da Costa in his unparalleled work on 'Medical Diagnosis' (1881)? He puts the matter thus:

'*Fatty Degeneration.*—Our power to recognise the fatty

changes during life has not kept pace with our power to recognise it after death. There is as yet no sign discovered by which we can positively say the dangerous disorganization of the muscular fibre of the heart is in progress. We may, however, suspect it if the signs of weak action of the heart—feeble impulse and ill-defined sounds—coexist with oppression, with tendency to coldness of the extremities, with a pulse preternaturally slow or permanently frequent and irregular, and be met in a person who is the subject of gout, or of a wasting disease, or is very intemperate, or has arrived at a time of life at which all the organs are prone to undergo decay. Sometimes more than a suspicion is warranted if, in addition to these, there be proof of atheromatous changes in the vessels, or of fatty degeneration elsewhere, such as an *arcus senilis;* or if it be ascertained that the patient suffers from severe pain across the upper part of the sternum, and from paroxysms of severe pain in the heart;* that he sighs frequently; that he is easily put out of breath; that his skin has a yellow, greasy look; that he is subject to syncope, or to seizures during which his respirations come to a standstill; and that he is liable to vertigo, or to be stricken down, with repeated attacks having the character of apoplexy, save that they are not followed by paralysis.'

Such, then, are the clinical features of fatty degeneration of the heart. It is found with other evidences of senile decay; and usually with failing hypertrophy.

The external appearance of the heart is as follows—the

* My own impression is that the agonizing pain of angina is due to the distension of the aortic root, rather than actually in the heart itself.

description having been written while frequenting the Pathological Institute of Vienna: 'The process does not pass on by continuity from one muscular bundle to another, or even from one fibril to another in a bundle, but is irregular; and side by side may be seen fibres in all conditions; from the still normal to the most advanced condition, with fibrils in various stages among them. The colour of the heart has become paler, and of a dirty yellow, or of a "dead-leaf" colour, somewhat unevenly though, and some parts are paler than others. The consistence is altered, the wall breaks down readily under the finger, and in advanced cases a finger may be pushed through the ventricular wall as through one or two thicknesses of wetted paper. The tissue is pliable, and tears with a sort of fracture, and crumbles down under the finger and thumb. It feels greasy and unlike normal heart-wall, and a warm knife passed through it looks oily. There is not necessarily any change in bulk, nor does the degenerate structure take on dilatation inevitably. The changes may be localized and confined to some particular portion, especially when due to disease of the coronary arteries.* This may give a mottled appearance to the heart; but, as commonly and more so, the heart is generally of this pale buff-colour, the blue coronary veins looking marked and distinct. The coronary arteries are not uncommonly tortuous, rigid, and atheromatous. The whole appearance in a well-marked case is that of degeneration, of partial death, or imperfectly renewed life, which is quite borne out by the microscopic appearances.' Sometimes the necrobiotic change is most

* In the great hospital of Vienna soft hearts were common as the result of pyrexia, puerperal and other.

marked under the endocardium, the external muscular layers being comparatively unaffected. 'To understand why the inner layers should especially suffer, we must remember that the innermost layers of the heart are farthest from the coronary arteries, and that they are broken up into *columnæ carneæ*, a condition which limits the directions of vascular supply to the attached ends of the columns, so that one perceives that these inner layers are under comparative difficulties in obtaining nourishment, which may explain the frequency with which this fatty degeneration occurs in and is limited to them.' (Wicks and Moxon's ' Pathological Anatomy.') When the heart as a whole is ill-nourished, the parts whose blood-supply is least abundant will naturally suffer most.

The condition of the walls of the coronary arteries and their branches is that of thickening, and diminution of the bore, or calibre; by which the blood-current through them is distinctly lessened. In the accompanying illustration

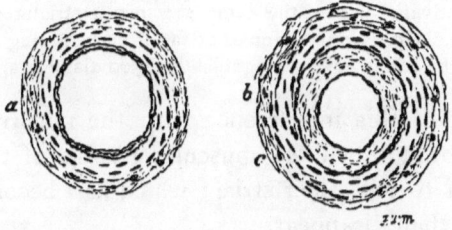

Fig. 17.—Transverse Section of two Small Arteries. Magnification, 300 diameters.

this is well seen. A is a normal artery; while B is an artery of the same size, which has undergone atheromatous change. It is easy to recognise at a glance the profound effect such a reduction of the blood-stream must exercise over the nutrition of the tissues; which derive their blood-supply from this altered vessel.

Muscular fibre is a somewhat loose combination of albumen and fat; and when its blood-supply is cut down, it breaks asunder into fat, and some nitrogenized body which finds its way into such blood-current as remains; or is absorbed in some way as yet unknown to us. The nitrogenized factor certainly gets away, leaving the fat. Such, indeed, is the physiological involution of the uterus after the expulsion of its contents in parturition. The

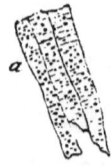 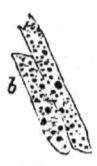

Fig. 18.—Fibres of Heart Muscle stained with Osmic Acid, from a case of fatty degeneration of the organ resulting from coronary sclerosis. *a*, Early stage, stain very indistinct, very fine molecules of fat. *b*, More advanced, from the same specimen; strictures in places altogether disappeared; granules of fat of very varying size stained black by the osmic acid. Magnification, 300 diameters.

first effect of this interference with the nutritive blood current is that the ill-fed muscular fibrillæ of the heart-wall begin to lose their striæ; which first become indistinct, and then disappear.

Such a muscular fibre in a comparatively early stage is seen in the fibre A in the accompanying illustration; while B is a fibre in a more advanced condition, which has indeed no longer any claim to be regarded as muscle. When the muscular structure of the heart is pervaded by this necrobiotic change it is softened, and has a tendency to yield; so that the once concentrically, or purely hypertrophied heart dilates; and a condition of mixed hypertrophy and dilatation is established; which distinctly lowers the heart's energies and power. It begins to fail; and is liable to be brought to a standstill in diastole by any demand upon it: whether effort, as hastening to catch a train; or by some vaso-motor disturbance, as arteriole spasm. A sudden rise in the blood-pressure in the arteries will overtax the weakened heart; and such doubtless is often the immediate cause of death in persons whose hearts are the seat of fatty degeneration. Or, maybe the feeble heart is arrested by some depressant toxic agent produced within the body itself. Dr. Lauder Brunton holds that the liver, as 'a porter at the gate,' is a filter which normally, and in health, arrests the malign and poisonous products of later digestion; and prevents their entering the general circulation. When the liver is impaired its action as a filter is reduced. In old-standing cases of vaso-renal change, and in advanced life, the liver as a filter fails in its duty; while the kidneys are the seat of extensive cirrhosis, and so are deficient in excretory power. No wonder then if, after an unusually hearty meal, death occurs in sleep from the accumulation of toxic products of digestion in the blood; acting as depressants of the heart's action.

But such sudden failure of the heart's action is rather the exception than the rule; the larger majority dying slowly of gradual failure of the heart's power, due to the degeneration and yielding of its muscular walls.

In the fatty degeneration of the muscular structure of the heart owing to the gradual occlusion of the coronary arteries and their branches, it is impossible to avoid some speculations as to the possible part syphilis may play in producing the result. Syphilis has a tendency to thicken the tunica intima, as is seen in p. 95, and the addition of a syphilitic element in the vaso-renal change may occlude the coronary circulation earlier than would be the case if the atheromatous changes had had no such ally. It is a matter well worth investigation by pathologists; and if this should ultimately turn out to be a well-founded surmise, it will add another to the many remote lesions which take their origin in syphilitic infection. At least the matter is one worthy of consideration, and investigation; whatever may be the conclusions arrived at by investigators. Clinically, one encounters not uncommonly cases of vaso-renal change, where there is also a history of syphilitic infection; but a wider field of observation is essential to determine how far a syphilitic element working in the atheromatous change, may antedate fatty degeneration of the heart-wall—from diminution of the lumen of the coronary arteries and their branches.

We can readily understand that the age at which atheroma of the coronary vessels becomes so established as to lead to the necrosis of the muscular structure of the heart, in a hale and otherwise healthy person with a good family history—is far in advance of that when death from cardiac failure occurs in a person of indifferent family history who has acquired syphilis early in life, and who has chronically indulged in alcoholic excess; and still more if to this be added times of privation and exposure. In this latter case fatty degeneration of the heart may occur at a comparatively early age.

Such, then, are the changes which go on in the structure of the heart itself, while the coronary arteries are tortuous, elongated, and of narrowed calibre from the adventitious growth in the tunica intima. The heart gradually wanes in power, the aorta becomes less elastic and more rigid; its recoil is diminished, and the general circulation is languid. As to the effect of the spread of this necrobiotic change throughout the muscular wall of the heart, the late Dr. Hayden wrote: 'This is no doubt most commonly the immediate cause of the fatal symptom when death proceeds from the heart in cases of hypertrophy from renal or valvular disease, or from atheroma of the aorta. When failure of the heart in a gouty subject is indicated by irregular action and palpitation, quickly followed by paroxysmal dyspnœa, and dropsical effusion becomes the signal of a 'break-up' of the constitution, as not unfrequently happens, fatty degeneration and dilatation of the heart will be found the last links of the chain, of which atheromatous change of the aorta, hypertrophy of the heart, and granular degeneration of the kidneys constitute the preceding links.'

As to the rate at which this disorganization of muscular fibre progresses, it may safely be said that it varies. In most cases the progress is slow—if it be possible to speak of a matter which cannot be seen, but only calculated—and probably is a matter of years. In other cases, again, an atheromatous tubercle of comparatively rapid growth blocks up one, or other coronary artery; and the progress of the case is comparatively rapid. Probably it is in these cases that sudden death takes place. In a case of sudden death under circumstances of excitement in an old gentleman in Hyde Park, the autopsy at St. George's Hospital revealed the fact that an atheromatous patch, so calcified as to resemble a scale of bone, had become loosened, and dislodged; and had closed the mouth of a coronary artery so completely that sudden death was the consequence. Some cases seem to march on at a very rapid rate. Such an one came under my notice two or three years ago. The patient was a well-known temperance lecturer; who had been lecturing with his usual vigour only a very few weeks before he fell down in the water-closet. His medical man and myself thought the *tout ensemble* of the case looked very like some rapid growth in connection with the coronary vessels; and warned him against any exertion, or effort. He kept very quiet for a week or two; and felt so well that he began to think that the gravity of the case was over-estimated. He went down to the City to arrange about a lecturing tour, came back, went into the water-closet; and died there. There was no post-mortem. But it may be said, and with truth, that in fatty degeneration sudden death is the exception rather than the rule. Perhaps, when fatty degeneration occurs with aortic incompetence, sudden death is frequent. Where one spot is gravely softened from occlusion of its own nutrient arterial twig, death may take place very suddenly from rupture.

More commonly by far, the heart becomes gradually impaired; and dilates as the softening progresses. Niemeyer, speaking of the prognosis of hypertrophy of the heart, wrote: 'The prognosis is often rendered more grave by the occurrence of a consecutive degeneration of the substance of the heart. With the transition from genuine to spurious hypertrophy, the picture changes, and many dangers arise.' As long as the nutrition of the large heart of vaso-renal change is maintained, all is well. But when the patient becomes less equal to effort; and exertion produces palpitation in a person who has not suffered from these symptoms previously; and a condition of hypertrophy blended with dilatation is found on physical examination, then we may fairly conclude that fatty degeneration is afoot throughout the muscular wall of the heart. To follow Niemeyer: 'The symptoms of dilatation supervening upon excentric hypertrophy, arising from endarteritis deformans, are essentially the same as those described above (Dilatation). It is often impossible to determine with which of these two forms we have to deal. This will not seem strange when we consider that endarteritis deformans does not occasion any derangement of the circulation as long as the heart remains in a state of excentric hypertrophy; and that the evidence of disease only becomes apparent after the hypertrophy has become spurious by secondary degeneration, and after its compensatory action has become imperfect. When we find by physical exploration that an old person suffering from cyanosis and dropsy has a dilated heart, that his superficial arteries are tortuous, pulsate visibly, and feel hard to the touch, the case is probably one of endarteritis deformans, with secondary degeneration of a heart which was once hypertrophied.'

As a matter of clinical experience, it is not by any means always easy on first seeing a patient advanced in years, and undergoing vaso-renal change, to determine whether it is a condition of blended hypertrophy with dilatation we have to deal with—a case where this combined condition has existed for some time; or it is one of a once purely hypertrophied heart beginning to yield, and dilate in consequence of the spread of fatty degeneration throughout its walls. Physical examination tells us the actual condition of hypertrophy combined with dilatation;

but no more. In order to make at all sure about the state of the heart-wall, the patient's history must be carefully elicited. If there be no new feature, and palpitation and shortness of breath on effort are old experiences; then probably the condition is one of old standing. If, on the other hand, the palpitation and dyspnoea on effort are new departures, and the bulk of urine has decreased steadily for some time; then the probability of the dilatation being of recent standing becomes very strong,—and the case is one of failing hypertrophy.

Arcus Senilis.—Evidence of degenerative change elsewhere is of great value in attempting to appraise the state of the muscular wall of a suspected heart. Such change in the eye has long been held of diagnostic and prognostic import; and is known as the arcus senilis. It consists of a crescent-shaped, or bow-like line at the junction of the cornea and the sclerotic; seen on raising the upper eyelid. Usually a like change can be detected under the lower eyelid. It takes place where the eye is covered from light by the lid. Gradually the horns approach each other, and then a ring is formed. This truly senile degenerative change must be distinguished from the ring of lime-salts which forms in many persons advancing in life. This bears some relation to the ring of bony plates found in some birds in this locality—the line of union of the cornea and the sclerotic. It forms a ring (*annulus*) from the first; is equally marked in the parts exposed to light, and those covered by the eyelids; it has a sharp, well-defined outline; while the central portion of the cornea is perfectly bright and clear. In the significant form the parts under the eyelid show the change most prominently, even when the horns of the bow (*arcus*) meet; the outline is hazy and blurred and ill-defined; while the cornea is hazy and cloudy from the presence of fat-granules throughout its structure. This condition of the eye—true arcus senilis—was well marked in the case of the woman who died from a dissecting aneurism (p. 97); and on microscopic examination of the heart-muscle considerable degeneration of the fibrillæ was found; some fibres normal, some losing their striæ, while others again were but a string of oil-beads within the sarcolemma; all lying side by side. (This bears out what Walshe said about extensive destruction of this kind having been found to exist, 'where death had occurred from unconnected disease of other organs.')

The Descent.—As said before, when degeneration of the heart-wall sets in, the case has passed 'the turn of the hill;' and commenced the descent. This descent, however, takes several directions. A few illustrative cases will best demonstrate this.

R. B., an old farmer of 70, was seen by me several times a good many years ago, but I was not his regular medical attendant. On one occasion he was propped up in bed, with his lower extremities very full of water, extending to his groins. His heart was weak, and its action irregular. He seemed very ill, and the look-out very dark. Nevertheless, eighteen months after this he was riding about upon his pony, looking after his farm, and as well as most men of his years. He was continuing well when I left the neighbourhood.

T. G., an old gentleman of 87, was sitting in his arm-chair, unable to lie in bed, with an aortic stenosis and much dropsy in both legs when first seen. He was put under treatment; and after being in his chair three weeks, he could lie down in bed. He got quite well of his dropsy; and resumed his usual habits, being apparently in good health for a man of his advanced years. Six months later, being sent for, he was dead before my arrival.

R. N., a retired farmer, aged 86, was also sitting in his chair, unable to lie down, with aortic stenosis and dropsy in both legs, when first seen. After a week or two he could take to his bed. In three or four months he was up and about, and was very active for his years. Two years later he had a return of his old symptoms, and died; but not under my care, and I can only speak generally about him.

J. H., aged 69, a brewery man, was admitted into the West London Hospital, January 25, 1877. He had dropsy up to his scrotum, and also in both arms. He had mitral insufficiency, while the aortic sound was muffled. Under treatment the bulk of urine rose, and the dropsy fell. On February 17 he left the hospital, and the house-surgeon wrote 'Discharged cured' against his name.

Dr. Cheyne put on record the following case: A gentleman, aged 60, a free-liver, had gout in his feet, and œdema in the ankles at night. He took a long walk, and suffered from palpitation after it. After that he failed generally; œdema increased in the limbs and extended to the lungs. He died of serous apoplexy. 'The heart was three times its natural size; its structure was extensively in "a soft, fatty condition." The valves were sound, and the aorta studded with steatomatous and earthy concretions.' (Taken from Stokes.)

The following case is taken from Dr. Basham's well-known work on 'Dropsy': J. L., æt. 50, was admitted into the Westminster Hospital for dropsy, extending up

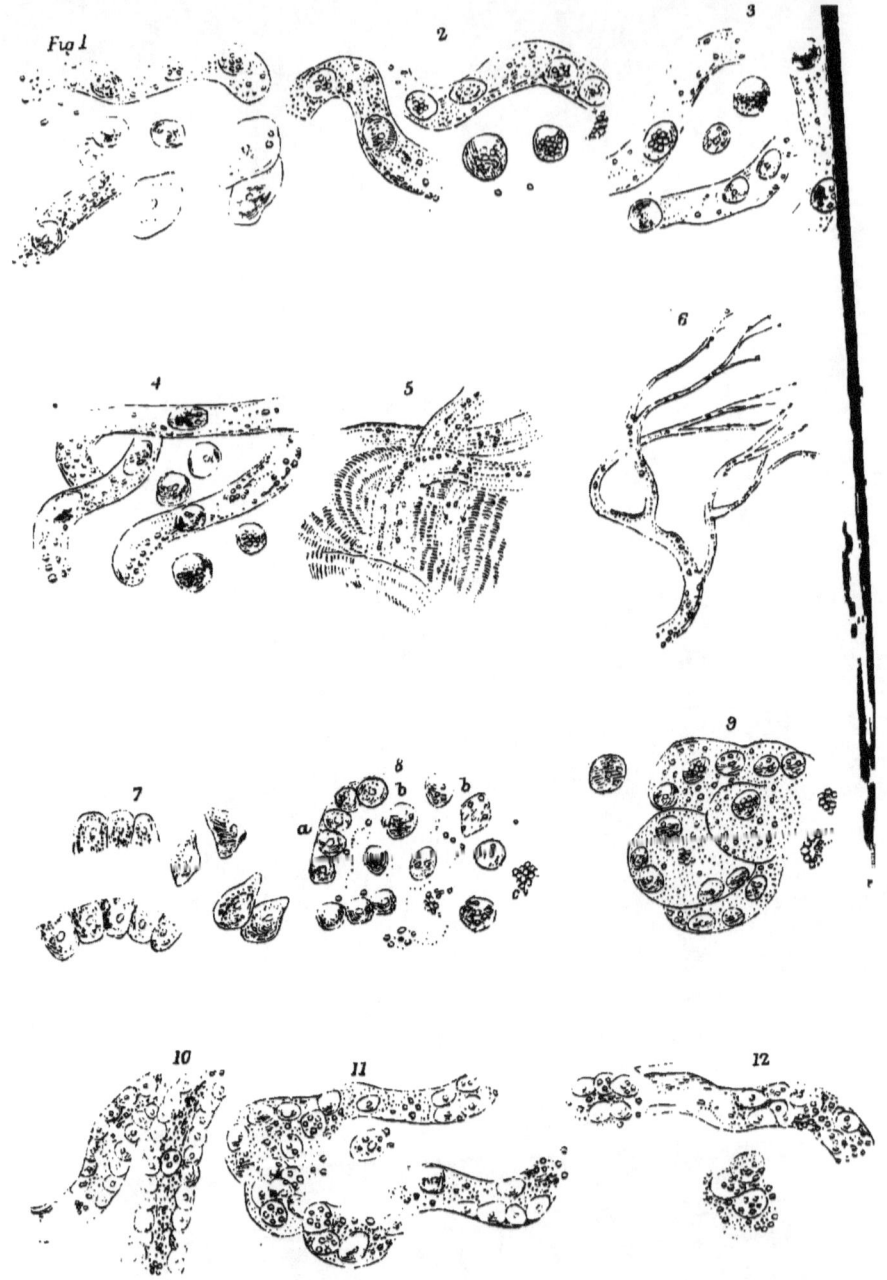

the legs to the genitalia. The aspect of the patient was characteristic of renal dropsy. After being six weeks under treatment, and feeling in better health than for months past, the anasarca having entirely subsided, he requested to be dismissed. Six months later he was again admitted, suffering in the same way, but worse. Again he improved to the point of talking about going out, when he was stricken with apoplexy, from which he died in four hours. 'An effusion of loosely coagulated blood was found in the substance of the pons Varolii, from which it passed behind the medulla oblongata. Blood was also extravasated into the fourth ventricle. The arteries at the base of the brain were opaque and rigid, the larger ones even quill-like; some small branches contiguous to the seat of the hæmorrhage were teased out with needles, and the microscopic appearance is represented in Fig. 6 of the accompanying plate. The heart was natural in size, the valves healthy; its muscular structure appeared paler than in a well-nourished heart. The kidneys were increased in size, lobulated, weighing severally eight and a half and nine ounces; in colour they were of a pale fleshy tint. Post-mortem observation showed that the heart-fibre and the arterial textures ultimately participated in this fatty decay. The muscular fibre of the left ventricles of the heart exhibited traces of fatty degeneration. The elastic coat of the artery is altered by this fatty deposit; it becomes easily torn and ruptured, and destructive hæmorrhage followed.'

Chronic Albuminuria.—Connected with degeneration of the arteries of the brain and heart-fibre.

1, 2, 3, and 4 are tube-casts; 5, a section of heart-fibre; 6, an arterial twig from the neighbourhood of the blood-clot; 7 and 8 are renal epithelium; 9, a portion of the cortical layer teased out with needles,; 10 is a straight tube from a cone; while 11 and 12 are tubes from the cortex. (This is a typical case of 'Bright's Disease' in its most exact sense.)

Not uncommonly chronic bronchitis is the prominent complaint. It is not at all uncommon to find an elderly person undergoing the vaso-renal change with a large heart, to have an attack of bronchitis which lingers in a chronic form; while mitral insufficiency develops with, of course, venous congestion behind it; in which the vessels of the bronchiæ share. There is, too, congestion of the pulmonic circulation. Before long some vesicular emphysema is added. A well-marked case of this kind came under my notice a little time ago at Bournemouth, whither the patient had wisely retired.

The accompanying plate 'was made from strongly-marked typical cases of emphysema, chronic bronchitis, and succeeding dropsy, by the late Dr. Basham. It gives a vivid conception of the state of the tissues where the closing scenes of the vaso-renal change are associated with bronchitis, and is highly instructive.'

Sir J. G., aged 56, came under my care with the following complications. He had œdema in both legs, a large heart beginning to yield; he had old-standing renal mischief, and a recent effusion into the right pleura. After a time a large infarct was developed in the left lung; and the respiration became so embarrassed that his medical attendant and myself saw no hope for him except by drawing off the fluid from the right chest, in order to give him breathing-space. Two pints and a half of fluid were drawn off with an aspirator, the lung coming down

Emphysema & Chronic Bronchitis.

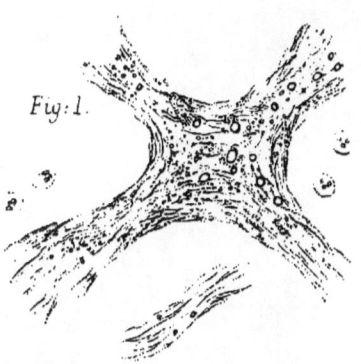

Fig: 1. Pulmonary tissue from emphysematous lung.

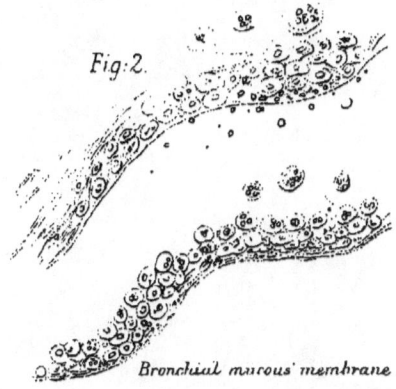

Fig: 2. Bronchial mucous membrane.

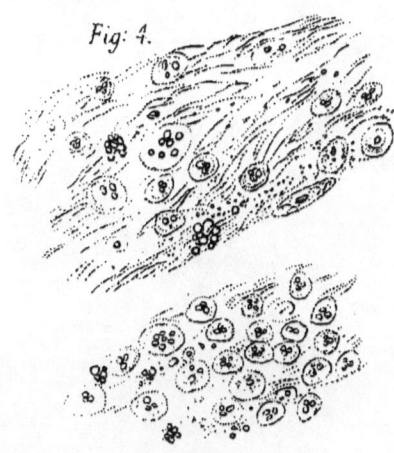

Fig: 3. Sputa Chronic Bronchitis with Dropsy.

Fig: 4. Sputa; Emphysema.

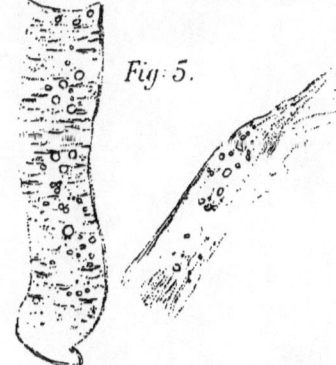

Fig: 5. Fatty condition of small artery adjoining emphysematous patch.

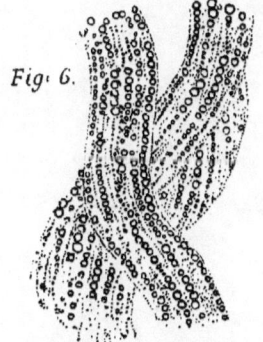

Fig: 6. From right ventricle. Heart Fibre.

From right auricle.

W.R. Basham. del. T. Way. lith.

in the most satisfactory manner; which relieved his breathing greatly, and enabled him to rally. Three weeks later death took place suddenly. There was no post-mortem.

Surg.-Gen. J. P., aged 72, was subject to attacks of angina pectoris vaso-motoria. He had the large heart and hard arteries of the vaso-renal change, with early commencing mischief in the aortic valves (insufficiency). The case progressed at a marked rate during the eighteen months he was under observation; death occurring from a severe fit of angina.

E. C., aged 60, had long manifested the indications of the vaso-renal change complicated with an impetiginous eczema, for which she was under the care of the late Sir Erasmus Wilson. Despite all he could do, she sank exhausted from the pain and itching of the skin affection; which extended over the whole of the body, except the face and hands.

These cases illustrate the various aspects assumed by the vaso-renal change when an advanced stage is reached. They might be multiplied indefinitely; but are sufficient to demonstrate the necessity for recognising the general condition underlying the prominent malady. The term, 'Bright's Disease' could hardly be applied to all these cases; in many of which the renal complications were latent, and the kidneys attracted no attention from beginning to end.

Venous Fulness.—The consequences of the failure of the heart's energy are felt in the venous system, and are called by the Germans Rückwirkung. Of course such are markedly developed where there is mitral disease, or any lung complication causing obstruction to the pulmonic circulation. (In such cases the pulmonary artery becomes the seat of the atheromatous change.) There may, or may not be any regurgitation at the tricuspid orifice; but such is commonly the case. Pulsation may be visible in the jugular veins; while the systole of the right ventricle is felt distinctly in the liver-pulsation, which is often very marked. Commonly the liver is enlarged; indeed, if it be not greatly enlarged, it is significant of cirrhosis, which prevents its swelling. The valveless portal circulation feels the venous congestion markedly. The spleen is enlarged. There is gastric catarrh giving a feeling of fulness; even when the viscus is empty. There is congestion of the vessels of the mucous lining of the bowels, with great evolution of gas; often termed 'heart-wind.' Sometimes the gas comes up from the stomach with roaring explosions. At other times gas is developed in the colon; and when the patient is already scant of breath from chronic bronchitis and emphysema, this causes great distress by preventing the descent of the diaphragm. Very frequently some wind gets pouched in the corner where the transverse passes into the descending colon; and much colicky pain is the result. This is often complained of by the patient as 'pain in the heart,' and is the more confidently referred to the heart if the wind compress the heart (through the thin diaphragm) with its elastic pressure, and cause the heart's action to be irregular. This may occur even if there be no valvular mischief present in the heart. Sometimes there is diarrhœa. Then there is often congestion of the hæmorrhoidal vessels, giving rise to piles, or rectal trouble, and much itching at the anus. Under such circumstances the prostate frequently enlarges. No uterine hæmorrhage, so far, has ever come under my notice as a sequel of heart-failure. Catarrh of the bladder, however, is found.

Fresh cloudy swelling occurs in the portions of the kidneys remaining sound, with growth of new connective tissue; the old mischief and the new often being con-

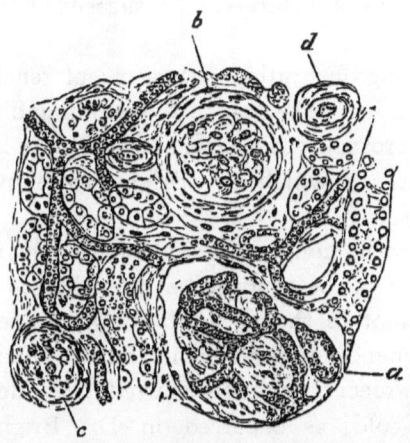

Fig. 19.—Section of the Cortex of the Kidney, from a case of vaso-renal change, in a man 56 years of age, with a syphilitic history. This shows an acute congestion occurring in interstitial nephritis. All the vessels are injected with blood, notably the vascular tuft of the glomerulus, *a*. Neither the glomerulus *b* nor *c* show any vessels ; probably because their function has been destroyed by the fibroid change which has occurred, leading to their contraction and degeneration. *d* is a small artery with thickened walls; particularly the intima is so increased, as almost to occlude the vessel. The epithelium of the tubules is undergoing parenchymatous degeneration. Magnification, 200 diameters.

spicuous by the contrast. Venous congestion is a certain provocative of albuminuria; and, sooner or later, albumen is present in the urine. At first it appears fitfully, but in time its presence is constant. At this point nothing could add to the gravity of the case; but the appearance of albumen is of sinister omen. There are times when some difficulty is experienced in estimating the gravity of albuminuria; but there is none here. Indeed it speaks only too plainly; and its accents are those of waning hope. Sometimes the patient makes a rally; but it is only of a temporary character,—to be succeeded by a graver condition.

The preceding illustration shows recent renal mischief occurring in kidneys the seat of old-standing disease. It shows the microscopic appearance of kidneys in the condition of those of which Dr. Bright has shown us the gross changes.

There is recent congestion amidst the old interstitial change. Such is the condition of the kidneys when new mischief is afoot in them; whether this be set up as a direct consequence of the venous congestion of a failing heart; or be recent nephritis started by a debauch, or exposure to cold, as occurred in Dr. Bright's cases. When such is the case the kidneys are swollen; and so larger than the small red granular kidney found when the vaso-renal change comes to an end without fresh kidney complications.

Dropsy, too, puts in its appearance. At first, perhaps the patient only notices some puffiness about the ankles at bedtime, all traces disappearing in the night; but the next evening the puffiness is again visible. An attack of bronchitis, or pneumonia may develop dropsy rapidly. A brisk purge, with diuretics, may give considerable relief; sometimes diarrhœa comes on of itself. In some persons, as in the case of R. B. (p. 165), the dropsy, after attaining very considerable proportions, passes away for a comparatively long time. This was also very conspicuous in a typical case of large white kidney at the Victoria Park Hospital. After the legs had been much swollen, but compatible with the patient's being able to walk, the case slowly yielded to treatment; and the woman was under care, off and on, for three years. She ultimately sank from uterine hæmorrhage at the change of life. More commonly the rally is of brief duration, the dropsy returning as the heart fails more and more; the urine becoming scantier and denser, and more highly charged with lithates. Effusion takes place into the serous sacs, which aggravates the distress; while effusion within the cranium is a not uncommon cause of death. Or œdema glottidis closes the scene.

Death.—Sudden death is found not uncommonly in the later, or even the final stages of the vaso-renal change. Sometimes it occurs in the midst of apparent health; and sudden death without enlargement (to any extent) of the decayed heart is one mode of ending in these cases. But far more commonly the case progresses with the measured march of a Gregorian chant to the bitter end; dropsy, effusion into the serous sacs, embarrassed breathing, and sodden limbs adding to the sufferer's discomfort, till life is indeed a burden. At last the patient dies; worn out with sheer exhaustion and carbonic acid poisoning.

Death presents itself in many terrible forms; but to see a person of good family history, and naturally good constitution, wear out with the phenomena of a failing heart, is one of the most painful of all sights which even the sick-room can afford.

Such, then, is the end of a change which often is visibly afoot twenty, thirty, or even forty years before its career is finally run. It presents us with many and varied phenomena, as we have seen. Probably all its manifestations are not yet within ken; and as time goes on and experience accumulates, other maladies may be found to be linked with it causally. The first departure physiologically is the reversion of the liver to the uric acid formation; and the presence of uric acid in excess in the blood provokes a growth of the lowly connective tissue—the wander-cells of the mesoblast—at the expense of the higher tissues, in the blood vessels and in the kidneys. The only other tissues directly affected are the two muscular ends of the circulation—the heart at the centre, and the arterioles at the periphery. But in time these become affected by the growth of connective tissue, which accumulates in the coronary arteries till the nutrition of the large heart fails; so that we see that in the end the connective tissue works the ruin of its muscular confederate.

But betwixt the first reversion to the uric acid formation and the failing hypertrophy—the beginning of the end—lies an immense area, covered by many maladies and much morbid action. Many of these have been hitherto regarded as independent maladies; but they can be seen to be really but parts and fragments of this vast vaso-renal change; and which in future will be described as such. Without some acquaintance with the whole morbid process it is impossible to rightly comprehend these various parts and fragments.

At least, they have become clearer to myself since being able to grasp the whole, and take a bird's-eye view of it; both as to their nature, their diagnosis and prognosis, and, further, their treatment. By tracing the pathological chain backwards, link by link, first one morbid action and then another come into view.

Getting back to the first beginnings, one is now able to detect the vaso-renal change at a very early period; and, consequently, to take measures to stay its course—to add to length of days, and avoid much discomfort. Feeling this in my own case, and realizing how my hands, as a practitioner, have been strengthened by this knowledge, my hope is that it may be equally valuable and useful to others.

CHAPTER VI.

PRACTICAL CONSIDERATIONS.

a. Insurance Office View of Life ; *b.* Surgical Aspect of Vaso-Renal Change ; *c.* Relations of Stomach and Liver ; *d.* Treatment— Dietary.

FROM the foregoing considerations it is abundantly clear that when a man 'gets the gout,' it is not something added to the healthy normal man; but something taken away from him. Indeed it is a *minus*, not a *plus*, quantity. The liver fails in its function of the urea formation; and, in its helplessness, reverts towards the uric acid formation of an earlier and lowlier form of life. It is, indeed, a species of involution—a turning back, or retrogression. In other words, it is the undoing of evolution. It is a loss of power, a falling back. The liver reverts to the uric acid formation; and the presence of urates in the blood-current is the portent of innumerable troubles, and ultimately grave structural changes. Vast and varied are the pathological outcomes of a blood surcharged with waste matters; which are themselves the outcome of albumen-metamorphosis. *Sublatâ causâ tollitur effectus.* If this uric acid formation can be arrested, put in abeyance, or to some extent rectified, then comfort and length of days are possible. Both will be in proportion to the influence that can be brought to bear upon the liver in its function of the metabolism of albuminoids, and in albumen-metamorphosis.

If the individual of acquired uric acid formation, or of inherited uric acid diathesis, persists in the meat-eating proclivities of his Norse ancestors, 'the beef and ale of Old England,' it is obvious to the meanest intellect, that matters must go on from bad to worse. When such an individual persuades himself first, and then protests that 'it makes no matter what he eats or drinks,' he is practising further indulgence—viz., a fallacy to cover his failings. It does matter; and a good deal, too!

Of course, it is not difficult to understand that a sudden repentance after long-existing excess will no more always work a cure in this liver-backsliding, than will repentance alter the character of a hardened sinner in all cases. If the individual goes on eating and drinking just what the palate pleases, the system will suffer. More and more uric acid will be formed; and further and more extensive injury will be done to the organism—bad going to worse. Attacks of acute gout may clear the system, as so many somatic thunderstorms. Secondary inflammations may achieve the same end, but carry with them a distinct danger to life. The kidneys must become further mutilated, or destroyed. The arteries will wax harder and more brittle, and so are more liable to rupture. Secondary valvular mischief is fostered in the enlarged heart. The spring is running down, as Dr. Goodhart so happily puts it.

To limit the albuminoid elements in the food is the first step to be taken. They should be brought down to the point of the actual requirements of the organism. The individual must have the issue placed plainly before him; and then he must make his choice. The alternative is not a pleasant one. Fourteen years of personal experience have taught the writer that it is nearly as irksome to avoid the gout as it is painful to have it. But the alternative is decidedly the safer practice. The inherited satisfaction at the sight of a huge joint of meat dies away as the conviction strengthens that it is not good food for one. Where common-sense rules, the appetites obey—in some other matters as well as the pleasures of the table. Self-control and the subordination of the appetites, like self-restraint in the matter of the passions, carries with it its own reward. 'Verily, of all carnal pleasures cometh satiety at the last,' said Kingsley's monk, when asked to sit after a prolonged experience of sitting on the seat of a boat. Lust leads to impotence. Indulgence at table ends in hepatic reversion. Gout in its Protean forms is the Nemesis of the meat-eater. 'Justice has a leaden foot, but an iron hand.' So may gout. Whatever statement the gouty man may put forward to excuse himself, the fact remains; and 'gout is the disease of those who will have it,' said Meade; and he was perfectly right.

A short story will put the matter in its most illustrative light. One day the writer met with Dr. Z., one of our most learned and honoured physicians. Discussing the matter of the uric acid formation, Dr. Z. said:

'You attach great importance to the dietary, Dr. F.?'

'Most assuredly,' was my reply. 'The avoidance of all albuminoids, especially the flesh of the larger animals, is the cardinal matter.'

His comment was: 'I am delighted to hear it. I am physician to an insurance office, and recently we had a proposal from a gouty man who sometimes had transient and intermittent glycosuria. Of course, our first impression was to reject the life summarily. However, it was decided to wait till the medical report came in. When it arrived, it was found to be signed by Dr. X. Now, I know Dr. X. very well, and also know that he would not let a patient of his eat, or drink too much; and on this consideration the life was ultimately accepted.'

The story is, indeed, not without its moral. It was just the same difference as with a man with a hernia. Well trussed, the life is a fair one; neglected, no office in the world would look at it. Allowed to eat and drink what he liked, and as much as he chose, the proposer was certain of rejection. In the hands of a judicious medical man possessed of much force of character, and able to influence a patient beneficially, the proposal was accepted. Nor can I regard the action taken as anything but sound common-sense. It is not likely that the proposer was accepted as a first-class life; but he was accepted. And that fact speaks for itself.

To reduce the dietary, as regards its proteids, to the capacity of the liver—in other words, to 'level down' to the liver—is the first step to be taken. This not only relieves the uric acid formation, but it gets rid of its untoward consequences. What did Dr. Murchison say before the Royal College of Physicians of London, in 1874?—' The day, I believe, will come when, with a more perfect knowledge than we now possess of the healthy functions, and of the signs of functional derangement of the liver, we shall be enabled to prevent, or to arrest at their commencement, many of the most serious maladies to which mankind are liable, and thereby to add another chapter to the volume of preventive medicine.'

Further intimacy with the liver and its derangements certainly places myself in the camp of Dr. Murchison; if it is impossible to go the length of Dr. Matthew Baillie. If persons of the uric acid formation could only realize the importance of adapting their dietary to the capacity of their livers, they would know how much they could add to their lives from an insurance point of view. That is one practical matter.

The Surgical Aspect of the Vaso-Renal Change.—For some time past surgeons have become fully alive to the importance of ascertaining the condition of the kidneys before operating on persons of middle age, or advanced life. Persons who have doubtful kidneys are best left alone, as the operation is pretty certain to end in disaster. For this end they have the urine carefully tested for albumen and sugar; and more recently a quantitative analysis of the urea is made. But even this last is but a mockery if one solitary examination only be made. The great variation of the urine from day to day is notorious in such persons. One examination may be perfectly misleading. The output of urea one day may be up to normal, or above it; but how about next day, or that day week? It may be far below the healthy average; and, what is more, may be far nearer the usual urine of the individual than that first examined.

Only while writing this chapter a neurotic lady of the uric acid formation brought some urine with her, as requested. Its specific gravity was 1025. From her account of herself, and the clinical facts of the case, instead of this being a fair average specimen of her urine, it was a most unfair sample. It represented the exception rather than the rule. It was, indeed, a dense, concentrated urine, following upon the heels of a large output of watery urine. Suppose the lady was a patient about to be operated upon, the urine, instead of speaking the truth, was lying as hard as an Oriental false-witness; and, instead of a guide, was utterly misleading.

The following case illustrates what is said above, and is distinctly instructive. In 1876 Mr. W. F. Teevan asked me to examine for him a man in St. Peter's Hospital who had come from South Africa to be operated upon for stone in the bladder. A tumour in the abdomen was the immediate cause of my being requested to see the case. 'I passed over the tumour as unimportant, but strongly advised no operation; stating it to be my opinion that the patient had chronic heart and kidney changes. This opinion was represented to the patient and his friends, but the patient begged to be operated upon. He was cut, and died. His kidneys were shrunken and contracted, till scarcely a sound piece could be found; one weighed one ounce and three quarters, the other three and a quarter. The heart weighed fifteen ounces, and was fatty. Such is the report of the post-mortem examination, which I did not see. The letter which the patient brought with him from the Cape, stated that repeated examination of the urine, chemically and by the microscope, gave no trace of albumen or casts; and on this negative evidence the kidneys were assumed to be sound. As the event proved, this opinion was utterly erroneous.' Such a case tells its own story, and carries with it its own moral. There was no carelessness whatever on anyone's part; the urine had been examined repeatedly, both chemically and with microscope. The medical attendant had conscientiously discharged his duty—up to his lights. Yet the result demonstrated, beyond any question or cavil, that the man was in the advanced stage of vaso-renal change; and fully vindicated my opinion that an operation should not be attempted. No doubt the letter from his medical attendant outweighed my opinion; and under identical circumstances would, with most surgeons; yet all the same the case furnishes a typical example of disproportionate trust in the renal secretion as an evidence of vaso-renal change.

In future, surgeons should have their patients thoroughly examined before deciding the question of operation: examined as a whole, and not in part. The urine characteristics are well enough; but they should be weighed in the balance in connection with the rest of the evidence. In a law court, in a complicated case all the witnesses are examined, and most of them cross-examined, before a decision is arrived at. The decision is not allowed to rest upon the statement of one solitary witness of doubtful character. And since the day of the late Dr. Mahomed the urine is no more than a doubtful character—as regards vaso-renal change. The condition of the vascular system taken as a whole is more worthy of trust than the urine. But no one portion of the widespread change should be taken exclusively; all, and everything should be weighed before a decision for, or against the vaso-renal change is made. Mr. Teevan's old gentleman put his faith in the result of the examination of the urine; and his life paid forfeit for his misplaced trust (probably the suffering he endured was not without effect upon his decision); and likely enough many another old man has lost his life from the false-witness borne by his renal secretion. When the Resident Medical Officer of the Leeds Public Dispensary, going round his wards with Mr. Pridgin Teale, we came to an old man whom he had cut for stone, whose appearance aroused my suspicions. An explanation was tendered to me that his water had been examined and found all right. On asking the man to put out his tongue, it revealed the brown hue of uræmia. The request was made that the kidneys should be examined at the autopsy. Both were found hard, granular, contracted and containing cysts.

When a surgeon suspects that a patient upon whom he is asked to operate (or desirous to operate) is the subject of vaso-renal change, or has diseased kidneys, he will do well to take some other matters into consideration than a mere specimen of the urine passed. That is another practical matter.

The Relations of the Stomach and Liver.—The mutual relations of the liver and stomach are of deep practical interest. Under certain circumstances the stomach is the protector of the liver; but at other times it is a fell tyrant. It may be well to dispose of the latter matter at once. Albuminoids are largely digested in the stomach; and the flesh of animals is often grateful to a stomach which resents vegetables and uncooked starch. Consequently, in order to avoid discomfort, a dietary consisting largely of flesh is adopted by many persons—much to the disturbance of the liver. Who does not know the sallow spare woman of the bilious diathesis, whose features tell of the sick headaches which she has endured; who lives upon a little lean meat, and a piece of dry bread with a little weak tea without milk or sugar: and declares that her stomach will only tolerate such food? She is bilious; and on such dietary she is likely to remain bilious. The stomach may be satisfied with such a dietary; but the liver is certainly embarrassed by it. There is a great deal too much albuminoid matter in the food for its comfort. It first of all provides biliousness with sick headaches, and a loathing of all food; which gives it a respite for the time at least. Later on in life the reversion to the uric acid formation is established. A number of cases have come under my notice where the craving of the stomach has led to embarrassment of the liver; and the person has become gouty through being a dyspeptic. One typical case of a bilious woman who lived on dry bread, a little lean meat and simple tea, rises up in my mind. She had been a sufferer for years; her face was indicative of often being bowed down with pain. After much exhortation she was induced to take to bread and butter, and milk puddings without eggs in them, and improved markedly; and ten years later was in much better health.

But far more frequently the stomach is the protector of the liver. A congenitally feeble, or what Dr. Budd denominated an 'insufficient liver' (a term endorsed by Dr. Murchison), is protected by a poor appetite, and a stomach easily offended. The one cuts down the amount of food taken; the other puts in its protest against any great bulk of food. Pain and suffering are the consequences of any indulgence at table. The simplest food only can be taken, and that too in very moderate quantities; or a penalty be exacted for any indiscretion. Any attempt to take a good meal, and then a long interval till the next meal, is fraught with discomfort. This is well seen in the case of the neurotic wife of a Norse husband. Oblivious of their physical differences, she will assimilate her meals to her liege lord's ways and customs. He has a good breakfast, a substantial luncheon, and in the evening an ample dinner. All very well for him, who has large viscera, and can take in large quantities of food at once, and then fast for hours—waiting for a good appetite to give zest to his enjoyment of the coming meal. But how about his wife with her small viscera? If she fast the same time, instead of a good appetite she has a headache, and is 'past' all food. He is like an express locomotive with a huge tender constructed to hold large quantities of fuel, so as to run long distances without stopping. She is like a yard-engine whose tender is little bigger than a coal-scuttle. If this engine set off for a long run, its tender would be empty before it got half-way. He enjoys a fast. It makes her ill. She must have small quantities of food at once, and at brief intervals. She can do very well with a little of something at eleven o'clock—an interlude betwixt breakfast and lunch. Some tea and bread and butter in the afternoon are acceptable. Kettledrum is a modern institution; and supplies a need with the modern neurotic of small digestive organs. Then such a lady can do very well with a little of something at bedtime; and in cold weather is all the better for some beef-tea and baked starch, or some milk and malted food in the small hours of the morning in cold weather. All put together it does not amount to much. But on such a dietary she is well and strong, *i.e.*, as well and strong as she can be.

   The confirmed dyspeptic in my experience is usually a neurotic of the uric acid formation. He growls at his stomach and complains of his dyspepsia, knowing what he feels; but still perfectly misunderstanding the real facts of the case. Such an one was Tom Carlyle—a typical example of the intellectual dyspeptic. Like all of this class he was irritable, and had a bad temper. 'That hag dyspepsia has got me bitted and bridled, and turned my living day into a waking nightmare,' he complained. As a matter of fact, dyspepsia could not 'bridle' him. It strove hard, but had little success. If it had not been for dyspepsia he would have gone to his grave ere middle age was reached. It was just this dyspepsia which he cursed so savagely that enabled him to make old bones. He was an ingrate, in fact.* Had he had a capable stomach, what would his liver have had to endure? There is a deep-rooted impression abroad—almost as strong as a religious conviction, that every individual ought to have four, or, at very least, three, good meals a day. That so much food ought to be eaten by everyone, no matter who; and many dyspeptics are sufferers for their devotion to their gastronomic duties. They do their duty to the cook's labours, and make themselves ill. A nineteenth century liver is 'the heir of all the ages'—the heir of the good and bad together. The Norseman's appetite is becoming a thing of the past. Their modern representatives cannot gorge themselves on pork, and swill beakers of ale with impunity, as did the Anglo-Saxon rovers, and the Norse vikings. Athelings, jarls, Norman barons, all fed, feasted and drank. Every man who could afford it kept a cook. Charles V. of Germany was a gross feeder, and retired from public life sodden with gout. He got his reward; and the picture of him drawn by Lothrop Motley is that of an ideal gouty man, who had at least

* He said of himself: 'I am a very unthankful, ill-conditioned, bilious, wayward son of Adam, I do suspect.'

'the sweet consciousness of guilt.' This is what Roger Ascham saw and said of him as Charles sat at the dinner of the Knights of the Golden Fleece, 'making his way steadily through sod beef, roast mutton, and baked hare, after which he fed well on a capon.' Nor did he forgot to drink with it all. 'He had his head in the glass five times as long as any of them, and drank no less at once than a quart of Rhine wine.' He deserved the gout: and he got it.

But the modern descendants of the old Norseman and the mediæval nobles, are degenerating as regards their eating capacities and their gastronomic performances. The blue blood carries uric acid on its current. The plebeian alderman who does not know who his grandfather was, is the capable feeder of to-day. The youth of *Punch*, who was down in the shires on hunting bent, and ordered up the round of boiled beef when the frost would not break (after an ample breakfast) was in a fair way to make the acquaintance of the gout. It is no mere accident of fashion that teetotalism and vegetarianism are in vogue at the present day; or that the brewer prepares a light beer rather than the strong ale of the past. Our railway-stations are placarded with the advertisements of prepared foods of all kinds. Why is all this? Because the English are losing their once famous powers of digestion. The child of to-day too often comes into the world with a feeble digestive apparatus, and is reared with difficulty. Often its mother cannot suckle it, and it is artificially fed, reared by hand; and its rearing taxes the best energies of all. And when, in its turn, it becomes a parent, its children will be still more difficult to rear.

And at this point something may be said about the neurotic dyspeptic of the uric acid formation as a parent. Those who reside in the country not unfrequently are prolific, having large families in some cases; and living to a good old age. But the urban neurotic, which means the bulk of these beings, is a failure as a parent. In my own experience she is frequently sterile; or has one child, a delicate little creature. If she have had more, they have perished in infancy or childhood. Mr. Cantlie has paid great attention to the degeneracy of Londoners, and on my inquiring of him what was the end of the degenerate Cockney race, his reply was to the effect that it usually succumbed to the maladies of childhood. He has observed a certain malformation of the lower jaw in London children of Cockney parentage; and this he has not seen either in the workhouses which he inspected, or in the dissecting-room at Charing Cross Hospital, in individuals over twenty-five; and very rarely over twenty, years of age. Probably the reversion to the Cymric type, which is so conspicuous in urban populations, is a degeneracy from the Norse type. And this degeneracy is evidently fostered by the circumstances of town-life.

The female neurotic dyspeptic is liable to trouble connected with her ovaries, in my experience. Commonly she has an irritable ovary, which is probably somewhat altered in structure; so that the full development of ova is thwarted. Certainly she has often a tender, swollen ovary, much aggravated at her monthly periods; and this must stand in some relation to her sterility. She certainly is a contrast in every way to the large woman—the typical mother of children. Small, active, energetic, unwearying; motherhood is not her vocation. Her one child, when she has one, is herself accentuated and more pronounced. She is an affectionate wife, a brilliant companion; but not the mother of sons.

Talking with an American lady, a perfect specimen of this type, she showed me her one little girl—a human gurnet; almost like an inverted cone, tapering from the head downwards—and asked my opinion of her; and what she would grow up into. My reply was to the effect that she would grow up 'a bundle of nerves, with shrunken viscera'—a not very hopeful opinion; but only too likely to be verified.

Such children should be sent away from the centres of civilization far into the country. Town-life is too stimulating for them. They are being civilized off the face of the earth. They are precocious. Puberty comes on while still immature. They perish early; old age comes upon them before the time of middle life. They frequently die of consumption. The lively, sprightly sylph, 'all up and down,' as her mother describes her; that is either elate in high spirits, or depressed with headache: she alternates rapidly. But the sword wears out the scabbard: the spirit exhausts its tenement of clay. Those die young whom the gods love: the fairy children, beautiful, precocious, charming, do not tarry long with us.

A still more degenerate being is the scrofulous or strumous child—the victim of joint disease, in whom the uric acid formation is so notorious. There is coarse-featured scrofula, and fine-featured struma; but the end is the same—viz., phthisis. The feeble digestive organs, the irritable stomach, and the insufficient liver cannot feed the system and the tissues. The loathing of fat interferes with the building up of healthy tissues; and tuberculosis is the result. Lugol (on Scrofula) has noted the effect of town-life upon the physique, reducing it towards scrofula ; which Laycock held was a reversion to a 'lowly ethnic form'—a term which quite meets the facts.

When the neurotic child is fairly well, it is desirable to let her have her own way about her food. Any attempt to feed her up, to raise her above what she is, only further embarrasses her liver. Cod-liver oil is, however, almost a necessary article of diet with her; being the only fat she can assimilate. It is not the best fat: but it is the only fat she can take. Milk puddings, and stewed fruit and cream, and milk with malted preparations, should form the staple of her food. Bread-and-butter puddings are excellent. If she loathe cod-liver oil, try her with toffee— the good old homely compound of butter and sugar, so acceptable to the young. Many a little neurotic child lives to bless me for pointing out to her mother the high food-value of toffee, and its utility as a means of supplying fat for the tissue-wants. Derision and surprise, of course, are the first results of the suggestion of toffee; because it has been looked at merely as a sweet—and not from its food-value point of view. At first it seems like harnessing Pegasus to a market-cart; but in reality it is an admirable combination of carbo-hydrates with hydro-carbons.

Such a child is a dainty, fastidious feeder ; a very bright child, but very difficult to feed. The bread and milk, or porridge and milk, so admirably adapted for the nursery fare, she loathes too often. Something tasty she can eat. Or the liquid fat of fried bacon she can digest; and ought to have. The time-honoured rules of the nursery, well suited to the children of the past, must be relaxed somewhat in her case. Compared to other children, she is a harebell in an onion-bed. She is the product of civilization; and her high-strung nervous system is linked with defective assimilative organs. Her liver is lacking in capacity; and exhibits a marked tendency to revert to the uric acid formation. To attempt to feed her up—under the impression that by so doing she will become robust, is a profound mistake. She cannot be fed up! An albuminoid dietary will not swell her muscles, and fill out her frame. It will only cause her liver to form more uric acid.

Poor man's gout is her ordained lot, *i.e.*, in some form or another. She often has palpitation, and has a comparatively large heart. The hypertrophied heart of anæmia is found in this class of girl. Only the other day a little mite appeared at the hospital, not yet thirteen, who has migraine with indigestion, acidity, flatulence, vesical irritability, and sediments in her urine; and on examination she certainly had a heart disproportionately large for her slight body. Her father and mother are robust healthy people; and the only cause why all their children are delicate seems to be the influence of town-life, and the fact that the mother bore the children rapidly.

The neurotic man has been given as delineated by Clifford Allbutt at p. 63, and it is a man of this type—a neurotic of the uric acid formation—whom Dr. Lauder Brunton had in his eye when he wrote: 'Feeling dull, weak, and melancholy, the sufferer now thinks he ought to take meat three times a day, and perhaps during the interval of his meals to take strong beef-tea, or perhaps a glass of milk and a nip of brandy. Yet in spite of all this he becomes weaker and more stupid, and more melancholy; and no wonder. He is simply further overtaxing his already overworked digestive organs. He is piling up fuel instead of removing ash, and choking the vital processes both in his digestive and nervous system. What he wants is not more nutriment, but a more rapid removal of waste; and the change upon the adoption of a proper system of treatment is, in many cases, most marked and satisfactory, both to the physician and the patient.' In the poor man's gout and the rich man's gout alike, the albuminoid elements of the food must be brought down to the capacity of the liver to deal with them. This is another practical matter.

Relations of Uric Acid to Acute Disease.—What relation the existence of uric acid in excess in the blood has to sundry acute diseases, is a matter upon which it is only possible to speculate. But some other matters than acute articular gout are coming within our vision. We have seen in this inquiry that indigestion, flatulence and acidity are as certainly due to uric acid in persons of the neurotic temperament; as is articular gout in persons of the arthritic temperament. New light is breaking in upon us. In all colds and catarrhs the practical man of a past generation commenced with a mercurial (usually combined with an antimonial) at bedtime, and some purgative medicine next morning; and then pursued the regular treatment. They called it ' clearing the system.' We are certainly all familiar with the pink, or red lithates of a common cold.

Then, again, there is another malady, which, like a common cold, is attributed to exposure; and which is marked by a large output of urates; and that is, acute rheumatism. The late Dr. Fuller, who was an authority on the subject, in the introduction to his well-known work, 'Rheumatism and Rheumatic Gout,' stoutly maintains that the starting-point is a *materies morbi* in the blood. 'And if the blood be altered in character it is practically the same, whether it contains matter foreign to the system, and altogether morbid in kind, or whether it contain an excess only of some material, a certain amount of which is compatible with health.' Further on he writes: 'The earliest and the most frequent victims of the disease, even when considering themselves in tolerable health, are apt to experience symptoms clearly denoting functional derangement. Though strong, perhaps, and equal to much bodily exertion, they are peculiarly sensitive to atmospheric vicissitudes, are prone to perspire, and their perspiration has a sour disagreeable odour, whilst the urine, though usually clear when passed, not unfrequently deposits, on cooling, a red brickdust sediment, a sediment of the lithates and lactates. So constantly are these symptoms associated with a tendency to rheumatism, that they have been recognised as indicative of a "rheumatic diathesis," or, in other words, of a case of constitution peculiarly prone to the incursion of rheumatism.' My own experience is in harmony with what is written here; and though acute rheumatism is not acute polyarthritic gout, it has sundry relations with the uric acid formation. Later on Dr. Fuller remarks very significantly, 'Moreover, rheumatism is so common among persons suffering from renal disease.' Lactic acid is held to be the particular poison of acute rheumatism; but its production has some association with a pre-existing state of the blood—though it is not yet possible to say what that association is.

Medicinal Treatment of the Uric Acid Formation.—This branches out into two lines—(1) the preventive treatment; and (2) the curative treatment. First come uric acid solvents. Sir Charles Scudamore used potash in acute gout; but with no further idea than the crude one that it was an 'antacid.' Dr. Garrod first pointed out that lithia and potash are true solvents of uric acid. Uric acid is highly soluble when combined with lithia, or potash; and if a specimen of urine containing lithates have some liquor potassii added to it, the sediments disappear and the fluid becomes clear. To wash the uric acid out of the body daily by some potash and lithia, taken every morning with a good draught of water—which is really treating the tissues to an alkaline bath—is an excellent practice; yielding the most satisfactory results, especially in florid persons of the Norse type. The Arab folk do not get on so well with this line of treatment, which is too depressant for them. Potash, as a muscle-poison, acts too potently upon them; and lithia is the uric acid solvent best suited to them. Some cannot stand lithia; and for these the acid phosphate of soda is indicated. Whatever the outcome of the uric acid formation—whether arthritic trouble, indigestion, migraine, or eczema—uric acid solvents rarely come amiss.

Colchicum is a notable remedy in gout; and certainly eases pain in acute articular gout. But the chronic resort to colchicum to keep off the gout is, in my opinion, a most pernicious practice — the person who does this becoming, in time, sodden with uric acid.

Hepatic stimulants are excellent. Whether it can be affirmed that these agents can raise the liver which is backsliding by reverting to the uric acid formation, exercising a direct action, is somewhat doubtful; but it must be admitted that the facts point in that direction. The good effects of blue-pill and a seidlitz powder upon a gouty, or a liverish state are well known by those who have experienced them.

Elsewhere ('The Diseases of Sedentary and Advanced Life') the writer has written: 'Next morning, after a good dinner, the gouty man is anxious about his property; his tenants are falling behind with their rents; his wife's health causes him the greatest uneasiness; while his eldest son's conduct is driving him to despair. After a blue-pill at night, and a seidlitz powder next morning, the outlook is rosier. He has one of the finest estates in the county, with first-class tenants upon it; his wife is hale, and in good health; while his eldest son is a young man of excellent promise, whose crop of wild oats is not going to be at all a lush one. And the altered state of mind is all brought about by getting rid of the excess of uric acid which so profoundly affected the mind, acting like a mental poison.'

The list of cases which Abernethy records where mercurials and cholagogues did so much good, point to cholæmia and lithæmia being the morbific states. A cholagogue like mercury is, however, not suited for habitual use. Rather should it be resorted to occasionally, and from time to time. Hepatic stimulants for habitual use are found in ipecacuanha, euonymin, iridin taraxacum, baptisin, and other vegetable products; while' the sulphate and phosphate of soda are also adapted for regular use (phosphate of soda is also a uric acid solvent). Some prefer the chloride of ammonia. Salicylate of soda is a potent cholagogue.

The vegetable hepatic stimulants seem to suit best the man of thin flank—the neurotic, or Arab type; and this is still more true of women. The florid gouty woman can take the uric acid solvents; but not so her neurotic sister. The two do not have trouble in the same way, and require different remedial measures. In both types, however, the regulation of the dietary is indispensable.

The Dietary.—The great source of uric acid seems to be the flesh of the larger animals. Upon this matter there is agreement among modern authorities on gout. Whether there is any failure in the primary digestion of proteids of this derivation in the alimentary canal; or the defect is strictly limited to the liver, is a matter not yet determined; nor likely to be for some time to come. But the empirical fact remains that a dietary which excludes this class of food suits best the person who is the subject of lithiasis in any of its numerous forms. The neurotic girl instinctively avoids animal food—by that same law of self-preservation which guides animals to avoid poisonous plants and seeds. White meats, as chicken and rabbit, are less harmful than brown meats (the cook sometimes classes pork and veal among the white meats, which is a mistake). Game and other birds, of course, are meats. Nor does the flesh of birds differ essentially from the flesh of beasts. The duck and goose are brown meats (but goose's liver is a capital form of food, albeit the English race has such prejudices against *pâté de foie gras*); so is the turkey. Fish is an excellent form of food, although it, too, is flesh. The white fishes are the most suitable for persons of the uric acid formation. Shell-fish of all kinds are good; and a lobster salad is a typical dish for the gouty man—who is perfectly confident about his digestive capacities. The oyster, with its huge liver, is good food.

Of all forms of albumen, caseine seems to possess the least tendency to go wrong in its metabolism in the body. A milk dietary often gives good results in cases of Bright's Disease. The Semitic Arab drinks milk and eats fruit; and so did the Hebrew of old, who travelled to 'a good land flowing with milk and honey.' Flocks and herds did the patriarchs of old possess in large quantity; but they did not eat of the flesh of them: as is shown by the story of the aged Isaac and the savoury meat. Corn and oil and wine, too, they delighted in; and cakes of fine flour and oil were in common use. But the flesh-pots of Egypt were not good; though apparently some murmurers looked back to them with regret.

Many modern neurotic Arabs of the uric acid formation are instinctively adopting the food-customs of the Arab and the Israelite; and with advantage. Pork and beef, the delight of the Norseman, are injurious to them; and 'the plain roast and boiled,' which the English housewife holds to be so wholesome, is just so much poison to them. The plain joint of Old England has played its part in history; and the Merry Monarch knighted the sirloin. And in the days of old beef was measured out by the yard. The Norse rovers found the pig a very convenient animal for their purposes. He had a good constitution, and a sea voyage did not disagree with him. His food requirements were simple; and his victuals could easily be stored beside him in a corner of the long war-keel. Other and later voyagers also realized his advantages at sea. Indeed, the Anglo-Saxon has had an attachment to the omnivorous brute since the dawn of history. Up to a very recent period fresh pork was the only unsalted meat of a country mansion in the winter months. The boiled round of salt-beef and the roast loin of pork were with many a country squire his staple dishes for a considerable portion of the year. Such food-customs continued for generations, and were accompanied by much physical exertion in the open air. Field sports, games, work on the land for long hours each day, enabled our old-world ancestors to endure their dietary, and to convert this nitrogenized waste into urea with considerable success. But times have changed. '*Tempora mutantur, et nos mutantur in illis!*'

With town-dwellers this is most certainly true. And the inhabitants of the town are now the larger half of the population. The tendency of town life to transform the Norse type into the neurotic type has been pointed out before. The man in town sets off to an office, usually an ill-ventilated room, at the hour when his grandfather set out on a hunting excursion. Active exertion in a bracing atmosphere enabled a liver to deal very differently with waste matters and *luxus consumption*, to what it can on the clerk's high stool, or the merchant's office chair in an impure atmosphere. But this matter is too often overlooked in the calculation. 'None of my people ever had the gout,' a town-dweller will exclaim with surprise to a medical man who makes the diagnosis of gout in his case. Certainly they lived under widely different circumstances. Their livers had the help of plenty of exercise and oxygen to carry on the urea formation. Their modern descendant's liver is handicapped by the rebreathed air of rooms and offices at a temperature of 70° Fahr.; and so reverts to the uric acid formation of the old Saurian in a tropical swamp. Is there anything to be surprised at in that? On all sides one hears the remark: 'How common gout is nowadays!' How could it very well be otherwise; when meat-eating is the practice of the wealthy, and the aspiration of the poor? Meat is the one means of acquiring health and strength in the eyes of many. Town life tending to the impairment of the digestive organs; the increasing consumption of meat by everyone that can afford it : are producing results which could easily be calculated *à priori*.

In connection with this matter of the consumption of animal food by town populations as compared to the rustic population, especially of past generations—we must have regard to the other fact of the defective oxidation of town life. In the country, much of the work, and the chief of the amusements are in the open air; while in towns it is very different. Work is carried on by an urban population in close confined rooms, with every crack and crevice closed against the income of air—in their dwellings the same is the case; while their amusements are indoor, with even a still more vitiated atmosphere. Their meateating customs encumber the liver; while defective oxygenation hampers it still further. If the kidneys become diseased, secondary to liver reversion, then it is amongst town populations we may look for the prevalence of vaso-renal change.

A large proportion of our population are discovering for themselves that vegetarianism and teetotalism suit them best. They are reverting in more ways than one. They are turning back to the Celto-Iberians as regards their digestive organs; and blindly and unreasoningly, but wisely, they also are reverting as to their food-customs. Farinaceous matters, fruits and milk are their dietary. The milk-pudding is supplanting the roast beef of Saxon

England. The water-caraffe is taking the place of the black-jack filled with foaming ale. Spiers and Pond have established a vegetarian restaurant. In the United States the bottle of aërated water with a basin of ice has taken the place of the wine bottle, till the American travellers in Europe have been studied by the innkeeper to see how far Boniface can compensate himself for their neglect of his wine list. The Australians drink tea, it is reported, to an extent that is injurious to them. Tea and coffee are the stimulant beverages for the modern neurotic; who recoils from the jugs of ale Queen Bess's maids of honour quaffed at their breakfast when they made eyes at Sir Francis Drake, and Sir Walter Raleigh won the heart of Elizabeth Throgmorton. For most of the wearers of 'the blue riband,' any such badge is perfectly superfluous. The slight physique, the comparatively large head, and thin flank mark them out quite sufficiently. The burly boor of old, fond of his corner at the tavern, is giving way to the slim clerk off to a temperance meeting, or the Young Men's Christian Association.

It is no use ignoring facts. The physician of to-day has to look beyond the walls of his study, or the wards of his hospital; and see how events are moving outside—if he wants to know what is going on around him. His father talked of 'the change of type in disease' as his lancet rusted in its case. Feeding up took the place of venesection; and now we are beginning to find out that feeding up is not a panacea for all evils. There are some livers which forbid that plan for resuscitating the race, and endowing the town-bred child of to-day with the constitution of a Viking of old—well-meant efforts doubtless; and if a certain place is paved with good intentions, graveyards contain many a mite who has succumbed to well-meant attempts to feed it up.

The modern physician has been looking too much inwards, towards his test-tube, and into his microscope; and too little outwards, in this matter. If he would read his history, and then take a turn through Madame Tussaud's, or the picture-galleries of Hampton Court Palace; he would find himself all the better able to grapple with the cases which come before him in his practice and walk in life. His grasp needs widening. Let him put what he sees through his microscope into contact with what he observes in Shoreditch, and its congeners; and with what he sees at the market-places of country towns on market-days. This would give him some solution of the fact that vaso-renal change is on the increase at the present time; and that reversion to the uric acid formation is part and parcel of a greater reversion. There is a reversion of a type, and that involves a reversion of food-customs—under penalty of disease and premature death. No wonder the spring has a tendency nowadays to run down too quickly!

If, too, it is the fact—as it certainly appears to be the fact—that the heightening of the nervous system goes hand-in-hand with waning power in the digestive organs; or, in other words, that the evolution of the nervous system of the epiblast involves the involution of the hypoblastic tissues; some other matters than the dietary ought to be looked to a little more. Is it quite fair on the part of parents to foster the tendency of their children to move in the neurotic direction; as is done on all sides? The country child grows thews and sinews, and develops into a stalwart adult. The town child goes to theatres and parties—all stimulating its nervous system and dwarfing its physique: it cannot eat its cake and have it. It is a man, or woman of the world at twelve; whether it is lounging in a carriage in Belgravia, or engaged in the interchange of chaff in Whitechapel. The town child has everything in action to favour precocity; in open defiance of the old adage, 'Soon ripe; soon rotten.' It is not an elegant expression—it is very plain, unvarnished old Saxon; but it is true.

In connection with the matter of early precocity—or, in other words, abbreviated infancy—the town child manifests a reversion to an earlier or lowlier ethnic form. Fiske, in his 'Cosmic Philosophy,' states, in speaking of infancy: 'In the human race it is much longer than in any other race of mammals; and it is much longer in the civilized man than in the savage.' He adds a foot-note: 'In this connection it is interesting to observe that the phenomena of infancy seem to be decidedly more marked in the anthropoid apes than in other non-human primates. At the age of one month the ourang-outang begins to learn to walk, holding on to convenient objects of support, like a human infant. Up to this time it lies on its back, tossing about, examining its hands and feet. A monkey at the same age has reached maturity, so far as locomotion and prehension are concerned.' But the ultimate development of the ourang-outang extends far beyond that of the lowlier monkey.

There is one other matter still to be considered in connection with the reaction of the epiblast upon the hypoblast; and the reversion of the congenitally insufficient liver to the uric acid formation: specially noteworthy in 'this madly striving age.' There is the haste to get rich.

It has been pointed out before that many of the migrainous dyspeptic ladies told of the long strivings of their fathers. The effect of mental strain upon the viscera has been considered. The fathers derange their digestive organs by their brain-labour. They beget children with incapable livers. To hard-working men this is a most serious matter. If their efforts to accumulate a competence exercise an injurious influence upon the physique of their progeny—and the facts certainly point in that direction—many men may pause, and slacken their haste to be rich. That the vaso-renal change is associated with fortune-making seems beyond all question.*

Vaso-renal change existed in the past: and probably such of the old sea-rovers as did not get killed in battle died of some outcome of it. Many a Teuton baron in his hall died some years before his time from his appreciation of the labours of his cook. But, all the same, it is widely on the increase amidst modern town-populations. Reversion to the primitive uric acid formation is part of a larger reversion. Change of type is carrying with it changes in food-customs—both as to meat and drink. The stalwart, fair-haired Norse folk are passing away, and the Celto-Iberians are getting their own again. And among the many other matters involved in this reversal is the pathological matter of vaso-renal change; while the inadequacy of the term 'Bright's Disease,' with our present acquaintance with the long and widespread pathological process is clearly apparent. If Richard Bright were alive at the present time, he would be the very last man to wish the term to be retained.

* This view is borne out by the frequent occurrence of diabetes and Bright's Disease among Jews.

L'ENVOI.

In the preface to this work the writer announced his intention to do his best to grapple with the widespread chronic pathological process to which he applies the term 'vaso-renal change;' hoping that some one may ultimately hit off a happier term. He has done his best: whatever amount of success he may, or may not have attained.

He feels that such a work is scarcely fitly performed by an outsider and adventurer: and that it should have been executed by some one engaged in medical teaching; and deputed by the Royal College of Physicians to watch and note the advances made in this department of medical science; and to report these to them from time to time in the form of lectures delivered before them, and issued to the world bearing their imprimatur. But as this has not been done, nor seems within measurable distance of being done: the advance of medical knowledge could not tarry till the venerable corporation in Pall Mall wakened up: and so the work had to be essayed by some one who would adventure it. This inactivity in the College is, however, quite consistent with its history. Its magnates have never given much encouragement to the advance of medical science; but have rather preferred

> 'To live and lie reclined
> On the hills like gods together, careless of mankind'

General Gordon held that the English colonies were first made by adventurers; and, when worth taking over, adopted by the Government. So it is in medicine. When some one has investigated a subject, and it has been accepted by the profession, it finds its way into the nomenclature of the College; the examining bodies accept it; and then the new recruit quietly takes its place alongside the veterans of the text-books—and becomes a formal part of the principles and practice of medicine of the future. Whether such will be the fate of the present venture time will tell.

The ideas propounded in these pages have in part, more or less, passed through the minds of others. As to the bulk of the clinical data, they have been gathered and collected by various and numerous observers: and all the writer can claim is to, more or less imperfectly, give order to chaos; and arrange many maladies, now standing in our text-books as separate entities, in their true place as parts of a great pathological whole. Possibly he may claim something for seeing disease in wider relationships than have hitherto been accorded to it. The work of Darwin has enabled him to see in the uric acid formation a reversion to the lower and earlier urinary products of the bird and reptile. Evolution involves involution: and in the return of town-populations to an earlier and lowlier ethnic form we see involution, or dissolution. Within this major involution, or reversion, we can recognise a minor reversion in the liver—which involves a long train of morbid sequences; albeit it may turn out to be really a long-sustained self-preservative, depurative action to rid the system of products which are injurious. There is the microcosm within the macrocosm; and the vaso-renal change, hight 'Chronic Bright's Disease,' may be Nature's means of weeding out degenerating beings; whose degeneration is due to the action and exigencies of town life.

The inquiry, too, brings us face to face with a matter which is beginning to exercise the minds of medical men who observe and think—viz., the relations of mental toil to visceral derangements. The impression is rapidly forming itself in many minds that long-sustained mental toil, or trouble, or vexation has an injurious effect upon the glandular apparatus of organic life. The effect produced upon the liver especially, *ex motu animi*, is being more widely recognised than hitherto; unless we go back to the ancients. Hepatic reversion to the uric acid formation injures the kidneys; while the blood, laden with waste matters, sets up many morbid changes elsewhere than in the kidneys.

But of the increase of Bright's Disease amidst us, and especially among town-dwellers, there can be no doubt. Town-dwellers lead a mentally more active life than country folks. The man who has made a fortune no longer retires to the country; but seeks in a town the advantages, and conveniences of civic civilization. The demands of the growing organism upon the nutritive powers are deranged by modern education. The epiblast makes demands upon the mesoblast at the expense of the hypoblast. The hard-working father begets children with insufficient livers which revert to the uric acid formation; with the consequence of interstitial nephritis. The returns of the insurance offices of the United States tell of a distinct increase in the amount of Bright's Disease amidst that enterprising and industrious people. There is no such thing as unalloyed good in this imperfect world; and vaso-renal change may be the Nemesis of ambition and of fortune-making.

That various parts of what is expounded here as to the vaso-renal change have passed through the minds of others, is seen by what now follows.

Dr. Blackall's work on 'The Nature and Cure of Dropsies' was referred to in the introductory chapter (p. 2), and is a very remarkable work. Dr. Quain drew my attention to it; saying how narrowly the old doctor had missed the vaso-renal change in its entirety. He observed the dropsy, the albuminuria, the occurrence of inflammations of the serous surfaces, and effusion into their sacs. He speculated as to the association of dropsy and albuminuria with renal changes; 'but whether this is merely accidental, and what relation it bears to the discharge of serum, must be left for future observation.' In discussing the causes of 'coagulable urine,' he goes on: 'But the most important and most fatal of all agents in producing this complaint still remains to be mentioned—an unsoundness of the digestive organs, which impairs the nourishment of the body, vitiates the blood, and gives vigour and operation to every other cause.' He continues in the next sentence: 'The free use of spirituous liquors greatly contributes to such an incurable taint, and, independently of that fact, has been thought by many physicians capable of exciting a true dropsy.' Then, in a discourse upon 'Angina Pectoris,' with which the work concludes—a very suggestive association—he notes the fact of palpitation occurring with it; and says: 'Palpitations are often rather a diminished than an increased action of the heart, the ineffectual and feeble efforts of a distressed organ.' He is troubled by the fact that angina is found with varying states of the heart—' in one instance emaciated, soft, and rotten; in another large, very hard, and strong; in a third the left ventricle was remarkably strong and thick.' As to its association with ossification of the coronary arteries, and disease of the walls of the aorta, this had been recognised long anterior to his time. Morgagni had observed the *cor magnum potius, et durum valde ac robustum.*

It is interesting to see how our modern acquaintance with the relations of syphilis to arterial degeneration had been foreseen darkly by Lancisi and Morgagni; while Scarpa held that 'aneurysm of the aorta is much more frequently produced by a slow, morbid degeneration of it than by violent exertions of the whole body, blows, or an increased impulse of the heart.'

Dr. Blackall was a thoughtful observer, and in connection with the vaso-renal change the following sentences are of high interest. In considering angina, he writes: ' Whatever differences of opinion may still subsist as to the constitutional cause which brings on changes of structure in the arterial system generally, and the particular varieties of these changes, it is, I believe, undeniable that gouty and rheumatic habits are most subject to the angina pectoris; but whether it is that ossification near the heart occurs more frequently in such habits, or that a slighter degree of them produces a great impression, may admit of doubt. This disorder makes a slow progress with indolent, rich, gouty persons; but I have seen it most rapidly destructive in a spare rheumatic habit, subject not indeed to the acute form of rheumatism, but to nightly pains without swelling.' Dr.

Blackall in this clearly recognises the two types—the Norse and the neurotic. He only needed the aid of the microscope to have anticipated the knowledge of a later day. It was no fault of his, but his misfortune; that this was denied him. What he could see, he did see.

The same division of individuals has suggested itself to another good observer, Sir Charles Scudamore; who had a very true eye for gout. He describes a class of gouty persons presenting more or fewer of the following phenomena, rather than articular gout: Headache, eructations which are sour and attended with a sense of heat; a craving appetite which does not become comfortably satisfied; oppression after a meal, with a painful sense of distension, and soreness of the whole epigastric region. Perhaps a slight meal causes a sense of fulness and distressing oppression; and the patient feels inflated, or, in his own words, *blown up*. The abdominal muscles are irritable, and convey the feeling of great rigidity on examination. To these may be added a furred tongue, with a viscid saliva, especially on rising; its taste is often remarkably saltish. There is much thirst. It is in these persons that the nervous system preponderates. This picture is pretty completely descriptive of the neurotic of the uric acid formation. Others, before the writer, have noted that the uric acid formation runs on different main lines, according to the diathesis and temperament of the individual; and it only required some close and continued observation to bring out the different points in each. And the division into the 'Norse' and 'Arab or Neurotic' for persons undergoing the vaso-renal change, has been accepted by a large number of good observers.

The effects of taxation of the brain as 'the organ of

mind' upon the glandular apparatus of the digestive organs, and the withering influence it exercises thereupon, have been recognised by many; and especially Dr. Charles Creighton in his interesting and suggestive work, 'Unconscious Memory in Disease.'

The late Dr. Thomas Laycock, the learned Professor of the Practice of Physic in the University of Edinburgh, in discussing the metastases of gout and rheumatism, held they were linked by a common origin of the affected tissues. For he wrote : ' Turning to the facts of embryological development or formation of these structures, facts at first sight are contradictory; for while the serous membranes, neurilemma, bone, muscle, and motor structures in general are developed from the serous layer of the embryo, and thus a community of origin and nutrition is manifest as to them; the heart and large vessels are derived from another primary tissue.' Professor Allen Thompson had not cleared up embryological development when Professor Laycock wrote the above sentence ; but the idea of disease affecting tissues on lines mapped out by the fœtal layers was clearly present to his mind.

Mr. Bland Sutton, in his recent and interesting work, 'An Introduction to General Pathology,' in speaking of neoplasms and their classifications, says : 'The most scientific basis on which to found a classification of neoplasms is undoubtedly an embryological one, leaving the histological details to determine the varieties. In this way they are divided into three great classes—those arising from tissues mesoblastic in origin ; those containing tissues formed from the epiblast and hypoblast ; and those composed of tissues derived from the three embryonic layers.' And further, he writes : 'The mode of classification is further sanctioned by the law known as the *specific nature of tissues*. For example, the cells of the epiblast never produce bones, neither do the cells of the mesoblast give rise to epithelium; and this specific character of the tissues arising from the three embryonic layers is maintained throughout the whole of life. Careful researches confirm this law.'

Then as to the heredity and antecedents of disease, my old friend Clifford Allbutt, F.R.S., of Leeds, has broached the subject in an interesting manner on several occasions. On my submitting to him an outline of the present essay, he writes me:

'The views contained in your letter interest me very much. When you ask me if I agree with them, I reply that I agree with you cordially in your main contention; and admire your method as a most fruitful one. Any difference we may find between ourselves in detail is of secondary importance. If I understand you aright you are possessed by the conviction which I began to preach in the *Medico-Chirurgical Review*, in 1863, that the only true nosology must be founded on the genetic relations of maladies and affections; and the true method of investigating disease is to trace its ramifications and variations through as many generations as possible. At the bedside I have tried for many years past to ascertain whether the woe of bringing special constitutional defects into society could be laid to the account of certain races of men; and whether our so-called tubercular, strumous, gouty habits, and so forth, are indications of qualities which belong to elements of racial admixture. The enormous difficulty of making even plausible approximative schedules of such kind has hitherto deterred me from formulating even sketches of such schedules in any public form. Moreover, great as is the inherent difficulty of the inquiry, its difficulty is enormously exaggerated by the reasonable and unreasonable prejudices of men.

'That men should shrink from exposing the defects and weaknesses of their family characters is natural and reasonable; and such reticence must be respected; on the other hand, we have to complain that we are misled ignorantly, stupidly, and even wilfully when searching into family histories on a legitimate mission. During the time of my association with the Collective Investigation Committee, I urged very strongly that complete stories of family tendencies should be sought for, as such stories are only known to the older and more trusted practitioners. But you make the more ambitious attempt to penetrate beyond the human family, and to investigate the phenomena of comparative biology; as I urged in my address to the Medical Section at Worcester in 1882; and beyond that to interrogate the phenomena of primary tissue-development, of which subject I lie in too much ignorance. If I may quote from my address, I will remind you of these words (*Rep. Brit. Med. Journal*, August 12, 1882): "We have to work out the genetic affinities of diseases themselves—their origin, parentage, and alliances, as well as their issues. We must seek out and define the laws or powers inherent in matter by which diseases appear, develop, vary, vanish, or prevail; and this not in the limited field of man alone, but far beyond it. The medical Darwin has yet to arise who will map out the evolution of disease. . . . We can have no complete therapeutics until the science of comparative nosology is in great measure constructed." A little later in the discourse I refer to "the long series of affections which belong to the order of lithiasis with its families and genera." But I shall weary you: and I will at once pass on to a point of view in which I may not find you my opponent, but which may not coincide with that which you have now undertaken to display. I am well known to be a fanatic for the nervous system. If I am then made somewhat blind to the power of other systems more or less equivalent, at any rate I deny that I can thus be led to overrate the paramount importance of the nervous system. The nervous system seems to me to be the register and organ of hereditary influences; and I believe that *in ultima ratione* nervous changes lie at the back of the arthritico-renal series and form their nexus. For instance, I do not think the relationship between lithæmia and chronic cardio-renal disease is direct, but indirect; that is, I believe they descend from common ancestry, if I may so speak—but do not generate either the other in the way of direct reproduction. That the cardio-renal changes are "the consequences of gout," as we see it crudely said, I do not believe. Cases meet us every day in which persons suffered from pronounced lithiasis for many decades, and died with their limbs crippled, having never shown any marked signs of cardio-renal change of the kind we have now in view. Atheromatous changes with anginiform symptoms are, indeed, more common in these than the simpler hypertrophy with reno-arterial fibrosis. On the other hand, we see this latter disease in persons who belong to gouty families, but who have shown either no common signs of lithiasis, or in whom these have quite a secondary position. No paper of mine, indeed, has received a more general adhesion, or been more regularly quoted by later writers, than that of 1876, in which I pointed out that "prolonged nervous strain was a potent cause of 'granular kidney.'" The common factor at the back of these diseases is some bias of the nervous system; and 'gout and granular kidney' are rather second cousins than parent and child.

'The relation of migraine to these maladies, again, is a very interesting question, and may prove to be a copula, or clue to many morbid kinships.

'Finally, I will not delay you longer to ask you to watch closely the relations of phthisis to these diseases. That phthisis and the more primary and malignant forms of "granular kidney" run together, I am pretty sure; and so does phthisis with acute rheumatism, the more atonic forms of gout and glycosuria—the last taking us directly to a definite nervous centre.

'I think, in a word, your forthcoming book is likely even to exceed your former ones in placing vividly before the practitioner of medicine the wider aspect of this subject; and in investing his work with more interest by your skill in grouping and colouring the details which labourers in other fields have accumulated in their more isolated way.'

This expression is highly gratifying to me; and the part played by the nervous system has not by any means been overlooked in these pages; and as our knowledge waxes, we may come to find it all that Dr. Allbutt claims for it. I venture to think that most readers will agree with him that such grouping of diseases makes the subject all the more interesting; both as to the whole, and as to individual factors.

Dr. Lauder Brunton, F.R.S., has approached the matter of reversion to the uric acid formation, from some experiments performed by himself and Dr. Cash, in a paper in the ' St. Bartholomew's Hospital Reports ' for the current year; and in writing to me he concludes: ' Our idea was that we should thus be able to find out the action of drugs on organisms which chiefly excrete urea, and on those which excrete uric acid. Our hypothesis was that we should find the action of drugs on the uric-acid-excreting organisms in all probability similar to those on gouty patients, especially during the paroxysm; but we did not think ourselves at liberty to publish our hypothesis until we had some experimental data. As you have, however, formulated very clearly and precisely the same hypothesis, it is no longer necessary for us to wait; and we have consequently mentioned our plan of research in our paper in the " St. Bartholomew's Hospital Report " for this year. I think that in all probability we might have waited for years before we had accumulated as much evidence as we wished before publishing our ideas; and it is very doubtful if we should ever have formulated the reversion idea so crisply as you have done.'

These matters and expressions of opinion tell that 'the thing is in the air.' In conversations with many of our leading teachers and authors, the writer has found others thinking along the same lines as himself; and recognising the widening relationships of what has been hitherto called ' Bright's Disease.' That the first departure is, as Hayles Walshe insisted, ' a change in the blood,' by which the vascular system and the kidneys alike, in time, become affected, is clearly recognised; and equally so that at the present and in the recent past the matter has been regarded too exclusively from the standpoint of the kidney and its secretion.

Of course, the present essay does not presume to claim to be exhaustive of the subject; but it is believed to pretty fairly represent the knowledge of to-day. To many readers it will be little more than their own ideas writ at some length. To others it will doubtless contain much that is unfamiliar to them; but, possibly, not on that account untrustworthy, or valueless. The writer certainly has felt himself the better practitioner as these views have formed themselves in his mind; and has learned to recognise the fact, that much of the disease of middle age and advanced life are truly depurative efforts on the part of the system: and he can only trust that his readers will have the same experience. Busy men in practice require books which will bring to them the information they feel they should like to possess; but have not the time to gather and work out for themselves. The writer therefore hopes this little work will meet a want;

'And where the vanguard halts to-day,
The rear will camp to-morrow.'

INDEX.

	PAGE
AFFECTIONS OF SKIN	71
Albuminuria	122
Aneurysm	97
Angina Pectoris	90
Apoplexy	99
Arab Type	58
Arterio-capillary Fibrosis	6
Articular Gout	48
Atheroma	94
Biliousness	69
Bronchitis Chronic	52
Change of Type	39, 59
Congestion of Diseased Kidney	173
Connective Tissue, Growth of	29
,, ,, in Kidney	33
Coronary Arteries, Changes in	157
Death, Slow	174
,, Sudden	173
Descent, The	165
Depurative Action of High Tension	25
,, ,, ,, ,, Dr. Mott on	26
Development of Localised Areas	14
Diagnosis of Gout	85

INDEX.

	PAGE
Dietary	196
Digestive Troubles	63
Dropsy	173
Eczema	54
Effects of Town Life	40
Embryonic Tissues	29
Emphysema	54
Epistaxis	102
Fatty Degeneration of the Heart	149
Fibrillæ, Fatty	158
Gout in England	16
Glycosuria	127
Heart, Fatty Degeneration of	149
„ Neuroses of	78
„ Secondary Valvulitis in	55
Insurance Office View of Life	176
Joint Gout	48
Kidney, Changes in	102
„ Congestion of Diseased	172
„ Normal	103
Lead	147
Liver Reversion	12
Mental Phenomena	74
Microscopic Appearances, History of	3
Migraine	72
Miliary Aneurysm	100
Nails in Gout	86
Neuroses, Cardiac	78
Neurotic Type	59

INDEX.

	PAGE
Palpitation	90
Poor Man's Gout	11
Psoriasis	55
Relations of Liver and Stomach	192
Rheumatism	51
Semeia of Gout	85
Skin Affections	71
Stage, Advanced	148
,, First	42
,, Middle	90
Stomach and Liver, Relations of	183
Surgical Aspect of Vaso-Renal Change	180
Teeth, Gouty	85
Traube's Views	5
Treatment	194
Tube Casts	116
Type, Arab	58
,, Norse	52
Urate of Soda, Deposits of	50
Uric Acid Formation, The	9
,, Relations to Acute Disease	192
Urine in Vaso-Renal Change	118
Valvulitis, Secondary	55
Vascular Changes	42
,, System, Changes in	144
Vaso-Motor Disturbance	17
Venous Fulness	170

FINIS.

Baillière, Tindall, & Cox, 20, King William Street, Strand.

Works by the same Author.

THE HEART AND ITS DISEASES: with their Treatment; including the Gouty Heart. Second Edition. Pp. 516. 16s.

CHRONIC BRONCHITIS: its Forms and their Treatment. Pp. 160. 4s. 6d.

INDIGESTION, BILIOUSNESS, AND GOUT.
> Part I.—INDIGESTION AND BILIOUSNESS. Second Edition. Pp. 320. 7s. 6d.
>
> Part II.—GOUT IN ITS PROTEAN ASPECTS. Pp. 300. 7s. 6d.

THE DISEASES OF SEDENTARY AND ADVANCED LIFE. Pp. 296. 7s. 6d.

THE PRACTITIONER'S HANDBOOK OF TREATMENT, OR THE PRINCIPLES OF THERAPEUTICS. Third Edition. Pp. 666. 18s.

MANUAL OF DIETETICS. Pp. 263. 9s. 6d.

FOOD FOR THE INVALID, THE CONVALESCENT, THE DYSPEPTIC, AND THE GOUTY. Second Edition. Pp. 167. 3s. 6d.